清华——MIT 城市设计联合课程作品集(2008-2014)
THU-MIT URBAN DESIGN JOINT STUDIO
STUDENTS' WORKS (2008-2014)

张杰 丹尼斯·法兰奇曼 毛其智 邵磊 编著
Zhang Jie　Dennis Frenchman　Mao Qizhi　Shao Lei

清华大学建筑学院　　SCHOOL OF ARCHITECTURE TSINGHUA UNIVERSITY
麻省理工学院建筑与规划学院　MIT SCHOOL OF ARCHITECTURE - PLANNING

中国建筑工业出版社

图书在版编目（CIP）数据

清华—MIT城市设计联合课程作品集（2008-2014）：汉英对照/张杰等编著.—北京：中国建筑工业出版社，2017.9
ISBN 978-7-112-21233-0

Ⅰ.①清… Ⅱ.①张… Ⅲ.①城市规划-建筑设计-作品集-中国-现代 Ⅳ.①TU984.2

中国版本图书馆CIP数据核字（2017）第225898号

责任编辑：杨　虹　尤凯曦
责任校对：李欣慰　李美娜

清华 — MIT 城市设计联合课程作品集（2008-2014）
张杰　丹尼斯·法兰奇曼　毛其智　邵磊　编著
*
中国建筑工业出版社出版、发行（北京海淀三里河路9号）
各地新华书店、建筑书店经销
北京嘉泰利德公司制版
北京方嘉彩色印刷有限责任公司印刷
*
开本：850×1168毫米　1/16　印张：11¾　字数：411千字
2017年9月第一版　2017年9月第一次印刷
定价：**78.00**元
ISBN 978−7−112−21233−0
（30876）

版权所有　翻印必究
如有印装质量问题，可寄本社退换
（邮政编码　100037）

前言
PREFACE

30 年前，在清华大学与麻省理工学院（MIT）老一辈教授的推动下，设立了清华-MIT 城市设计联合课程。在过去的 30 年中，这个课程两年举办一次，从未间断。先后有近百名来自于两校的教师参与了教学，有来自美国、中国等几十个国家的 400 余名学生选修了课程。清华-MIT 城市设计联合课程体现了中国改革开放以来中美一流大学在教学育人方面的深入合作，也见证了中国城市天翻地覆的变化与城市规划、建筑学学科与专业教育的发展。

在前 20 年里，课程教学主要聚焦于北京的城市问题，教学主题涉及旧城保护与更新、城市边缘地带的改造与景观环境治理、城中村改造与社区发展、轻轨建设与城市规划等。2006 年在清华-MIT 北京城市设计联合课程 20 周年之际，我们曾在北京城市规划展览馆举办了一次大型教学成果展，展览内容后由清华大学出版社出版。

从 2007 至今近十年的时间里，双方共举办了五次城市设计联合课程，选题转向城市可持续发展这一重大议题。2008 年在北京举办奥运会之前，课程以首钢搬迁为对象，探讨首钢工业区再利用问题，站在了国内大型老旧工业区改造与利用领域的前沿。此后，课程面向日益严重的全球环境与气候问题，将清华大学与 MIT 在低碳城市领域的研究成果与课程教学密切结合，探索低碳城市设计的理念与方法。课程选择了济南、太原等冬冷夏热地区开展研究与设计工作。通过科研与设计互动的方式，采集了大量一手数据，合作团队初步建立了低碳城市计算模型，并借助模型指导学生优化低碳城市的规划设计。2016 年 6 月正值清华-MIT 城市设计联合课程 30 周年之际，双方又在北京城市规划展览馆的协助下成功举办了这一阶段的学生作业展览，较完整地反映了这一时期的教学研究的探索过程。与这些设计成果相平行，双方师生还完成、发表了大量的学术论文。所有这些都为研究性设计的教学与科研工作开拓了崭新模式。

为纪念清华大学建筑学院成立 70 周年，在清华大学建筑学院的资助下，我们特将近十年的课程学生作业整理出版。

在清华-MIT 城市设计联合课程近十年教学成果出版之际，我们要感谢清华大学建筑学院、美国能源基金会对课程的资助，感谢北京市规划委员会、济南市规划局、太原市城乡规划局对课程的大力支持。

The THU-MIT Joint Studio of Urban Design was set up with the effort of professors from both universities 30 years ago. In the past 30 years, the studio has been held biennially and never been suspended. Nearly 100 teachers from both universities have devoted in the studio, along with more than 400 students from tens of different countries around the world including the United States and China. The Joint Studio reflects the great cooperation between the top universities of the United States and China in education ever since the reform and open of China It also witnesses the unprecedented change of Chinese cities along with the great progress of the education of urban planning and architecture.

In the first 20 years, the studio mostly focused on the urban issues in Beijing. The subject involved the reservation and renovation of old city, the reform of peripheral region of city with the landscape rehabilitation, the reform of urban villages and development of communities, and the construction of light railway and urban planning. On the 20th anniversary of THU-MIT Joint Studio in 2006, we held a grand exhibition of the teaching result in the Beijing Planning Exhibition Hall. The exhibition has been published by Tsinghua University Press.

In nearly ten years since 2007, we have held 5 joint studio of urban design, with the topic shifting to sustainable development of cities. Before the Beijing Olympics in 2008, we focused on the relocation of the Capital Steel Company, discussing the reutilize of the old industrial park, which put forward the research of renovation and reutilization of old industrial district in China. Since then, to face the growing seriousness of global environmental and climate issues, we seek to explore idea and method of Low-Carbon urban design, by inosculating the research and teaching result in the field of Low-Carbon city with both universities. We chose cities like Jinan and Taiyuan to be sites of our research where is cold in winter and hot in summer. By establishing interaction between technology and design, the joint groups have gathered abundant first-hand data, setting up initially the model of Low-Carbon city for calculation, instructing the planning and design to optimize Low-Carbon city. In June 2016, the 30th anniversary of the THU-MIT Joint Studio of Urban Design, we also successfully held an exhibition of the students' works in this phase with the assistance of Beijing Urban Planning Exhibition Hall, where the process of teaching and research exploration is reflected. Meanwhile, plenty of academic papers are also published by the teachers and students in the Joint Studio. All these efforts are helping to establish brand new pattern of the academic and scientific research and design.

To commemorate the 70th founding anniversary and subsidized by School of Architecture of Tsinghua University, we specially edit and publish the students' works for the past decade.

On the publication of THU-MIT Joint Studio of Urban Design, we'd like to thank School of Architecture of Tsinghua University, and USEF for the investment, and Beijing Municipal Planning Commission, Jinan City Planning Bureau, Taiyuan City Planning Bureau for the substantial support.

目录
CONTENTS

2008
首钢 SHOUGANG CAPITAL STEEL PLANT — 001
重组首钢：衍生公共空间 RE-INTEGRATING INDUSTRY: CIVIC LAYERING — 003
绿色首钢：全球可持续发展研究中心 GREENING THE FACTORY: A GLOBAL SUSTAINABILITY INSTITUTE IN BEIJING — 014
都市农田 URBAN AGRICULTURE: A VERTICALLY-INTEGRATED APPROACH TO SHOUGANG'S SOCIAL, ECONOMIC AND PHYSICAL REVITALIZATION — 020

2010
济南 居住区 清洁能源城市 JINAN NEIGHBORHOOD CLEAN ENERGY CITIES — 027
社区生活，绿色城市 COMMUNITY LIFE, GREEN CITY — 029
活力社区 VIBRANT COMMUNITY — 041
低碳生活 LOW-CARBON LIFESTYLE — 050

2011
济南 西部新城 清洁能源城市 JINAN NEW TOWN CLEAN ENERGY CITIES — 063
清洁能源城市 CLEAN ENERGY CITY — 065
空中城市 STALAGMITE CITY — 069
清洁能源城市 CLEAN ENERGY CITY — 073

2012
济南 城中村 清洁能源城市 JINAN URBAN VILLAGE CLEAN ENERGY CITIES — 079
农业—文化—生活实验室 AGRI/CULTURE LIVING LAB — 081
知识庭院：人与自然、人与人之间和谐交流的场所 COURTYARDS OF KNOWLEDGE: CONNECTING PEOPLE, PLACES AND IDEAS — 091
活力新城 VILLAGE FOR VITALITY — 101
济南西部新城 OASIS NEIGHBORHOOD — 111
生产的城市 THE PRODUCTIVE CITY — 121

2014
太原 城中村 清洁能源城市 TAIYUAN URBAN VILLAGE CLEAN ENERGY CITIES — 131
五种 FIVE SEEDS — 133
社会形式 SOCIAL PROFORMA — 139
解 DISTRIBUTED MILLENNIUM — 146
原之线 URBAN TREADS — 154
韵转未来 FLEXIBLE FUTURES — 161
街道主义 STREETISM — 168
路径 LIVE/WORK/PLAY PATHS — 176

致谢 ACKNOWLEDGEMENTS — 183

2008

首钢

SHOUGANG
CAPITAL STEEL
PLANT

首钢搬迁是北京城市经济结构调整、环境污染治理的重大战略决策。首钢是中国十大钢铁企业之一，也是唯一立足于首都北京的钢铁企业，创办于1919年，发展高峰时期员工多达26万人，120多家下属单位和分厂，曾经是世界上规模最大的钢铁企业，在北京过去几十年的城市发展中，扮演着举足轻重的角色。然而，伴随着北京空气污染的日益加重、水资源的严重短缺、运输物流成本的不断增加以及北京城市发展战略的转型，首钢作为大型钢铁制造业从北京搬迁出去成为必然。搬迁后的首钢，面临工业用地的再开发再利用的难题，如何从宏观城市结构到微观建筑形态，从经济发展、社会文化、生态环境等方面与城市有机整合，实现工业用地的活化与复兴，是对这个地区展开规划设计研讨的重要目的。

The relocation of Shougang is a significant strategy in terms of adjustment to economic structure and attainment to curbing environmental pollution. As one of the most 10 successful steel production companies and the only one located in Beijing, Shougang has been the largest steel company in scale of operation among the whole world since the foundation in 1919. There are almost 260 thousand of workers and more than120 subsidiary factories of Shougang in its peak time. With no doubt that it has played a decisive role in urban development of Beijing during the past decades. However, with the increase of air pollution, a serious shortage of water resources, the increase of transport costs in Beijing, as well as the transformation of Beijing's development strategy, it is inevitable for Shougang as a large steel manufacturing to relocate from Beijing. After the relocation of Shougang, Beijing will face the problems of redevelopment and reuse of industrial land. How to organically integrate the site with the city and achieve activiting and reviving of industrial land, from macro urban structure to micro architectural form, based on economic development, social culture and ecological environment, is an important aim to carry out the discussion of planning and design on this site.

重组首钢：衍生公共空间
RE-INTEGRATING INDUSTRY: CIVIC LAYERING

Claire Abrahamse, Christine Outram, Josh Fiala, Zhai Wensi, LiYe

1 基地位置 Site Location
2 现状要素 Status Quo
3 公共空间衍生策略 Civic Layering Strategy

为了将首钢地区重新编织到城市网络中，并建立一种能够强烈回应城市发展的公共网络，我们提出了以下一系列促使地段与城市联动发展的策略：

1. 阶段性开发模式+土地复垦；
2. 分区分阶段地开发公共空间网络，使其在地段的阶段性发展中得以延续和加强，并由此建立社区中心以及城市的结构性联系；
3. 保留并重新诠释地段在城市中的工业角色，以一种动态而有意义的方式，保护特定的现状空间、结构以及技术；
4. 重新建立首钢地区开放空间与北京城市结构的联系，使河岸、山地与绿色空间回归城市。

• Economic value, through developing in a series of iterative layers or stages and through the remediation of certain areas of the site;
• Social value, through prioritizing the development of a dignified public-space network, which would remain and be reinforced through itera- tive development layerings of the site, thus creat- ing community centers and urban connections;
• Cultural value, through the retaining and reinterpretation of the industrial role of the site within Beijing, thus preserving certain existing spaces, structures and skills in a dynamic and meaningful way;
• Environmental value, through the linking of Shougang's vast riverfront, mountain areas and green spaces back into the public structure of Beijing.

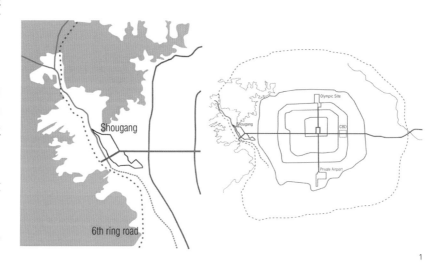

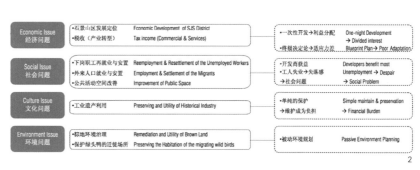

我们提出以"创新工业世博会"作为初始策略，促使整个演进过程的开展。这一策略具有很强的可行性，并将发挥触媒作用带动整个地区的发展。同时，这一工业世博会的新形式将使工业在城市中的潜在价值得到重新审视，并加强地段所具有的传统与创新工业的双重特性。

In determining a "first move" catalyst layer, the highly-imageable notion of an exposition was drawn upon. A new type of industrial exposition is proposed for the site as an instrument for re-examining the potential role of industry in Beijing and solidifying the identity of the site as a space of both industrial innovation and tradition.

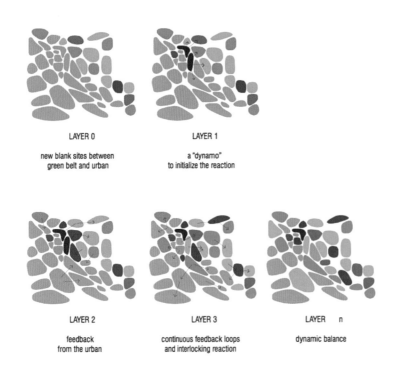

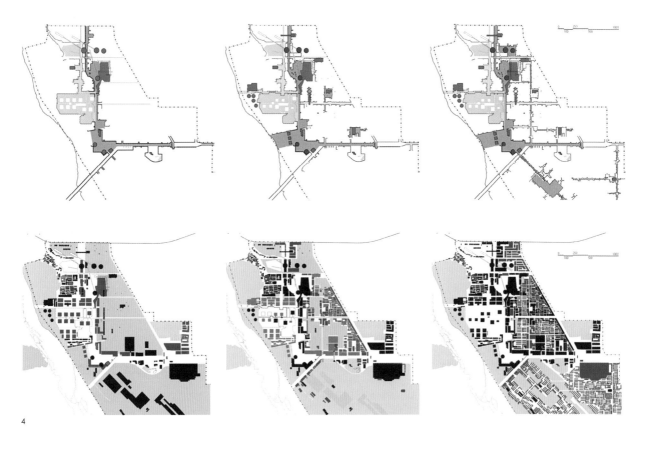

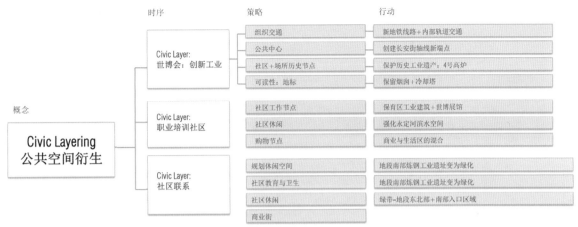

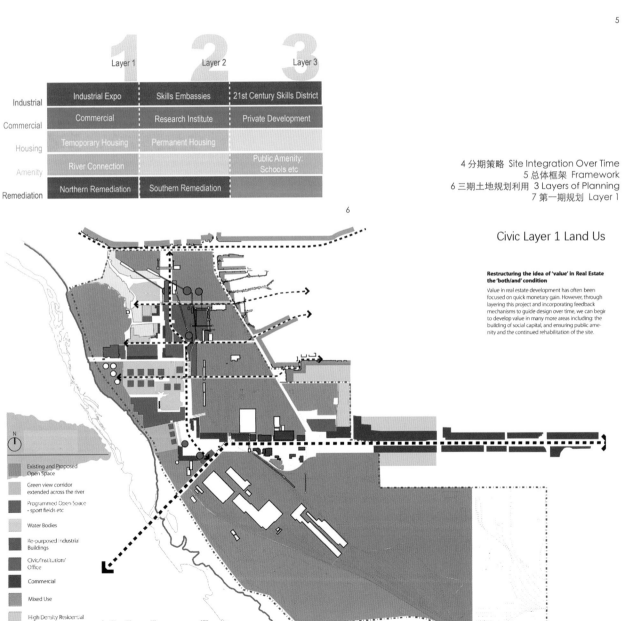

4 分期策略 Site Integration Over Time
5 总体框架 Framework
6 三期土地规划利用 3 Layers of Planning
7 第一期规划 Layer 1

Civic Layer 1 Land Us

Restructuring the idea of 'value' in Real Estate the 'both/and' condition

Value in real estate development has often been focused on quick monetary gain. However, through layering this project and incorporating feedback mechanisms to guide design over time, we can begin to develop value in many more areas including: the building of social capital, and ensuring public amenity and the continued rehabilitation of the site.

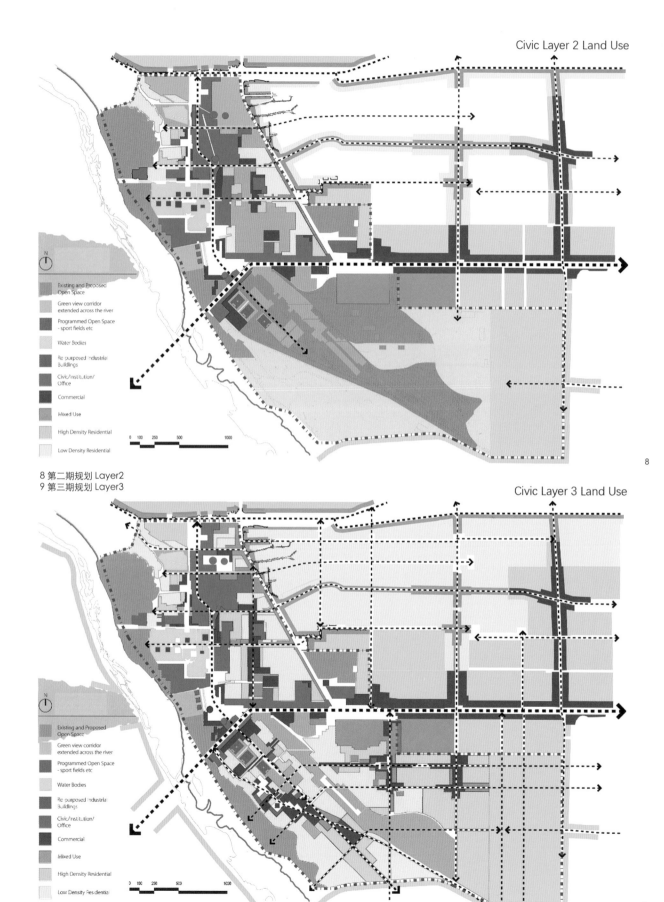

8 第二期规划 Layer2
9 第三期规划 Layer3

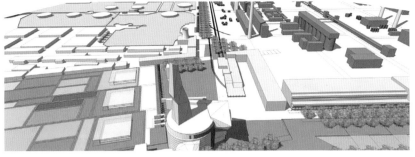

世界创新工业博览会
博览会不仅沟通了场地,同时创造了东西轴线的有力的结尾。根据我们的组织逻辑,其结果是:创造性的结尾和轴线的转移。

World Innovation Industrial Exposition
The Expo program gives us the ability to not only create access and mobility connections across the site but also to improve public amenity and make a bold statement at the end of the East-West Axis. In relation to our organizing principles this has resulted in:
The conceptual finishing of the east-west axis at our site and diverting the axis in line with the existing factory grid to the south of the site for it to be developed.

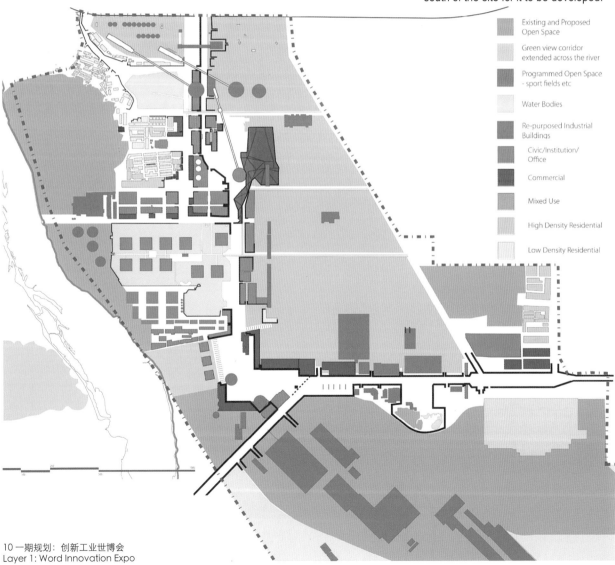

Existing and Proposed Open Space
Green view corridor extended across the river
Programmed Open Space - sport fields etc
Water Bodies
Re-purposed Industrial Buildings
Civic/Institution/Office
Commercial
Mixed Use
High Density Residential
Low Density Residential

10 一期规划:创新工业世博会
Layer 1: Word Innovation Expo

它的中心是一个新的"鼓楼"，工业遗迹重新组织在公共空间里。

A layering of themed thresholds leading up to the termination point of the East-West Axis, which is centered on a new "drum tower" – an industrial relic reinterpreted and adapted to reveal the develop- ment of the site in a public forum.

北部新的大门通向最重要的1号炉。

A new gateway in the north of the site that directs traffic past furnace no.1 (the most prominent furnace)

4号炉的内部中心将成为工业档案和数据的中心。

An internal center at Furnace No. 4, which is adapted in order to accommodate an industrial archive and digital data bank for new processes on the site.

新的地铁、轻轨、公交系统。

New subway and rail stations, as well as internal light-rail and bus systems.

初步整治土地，增加土地价值。

Initial remediation of the site so that the land. This will also increase the value of the land when it comes time for it to be developed.

联通水系，引入灌溉工程。

Revealing the water links between the two lake areas and leading water from the lakes into the remediation areas in order to irrigate the remediation.

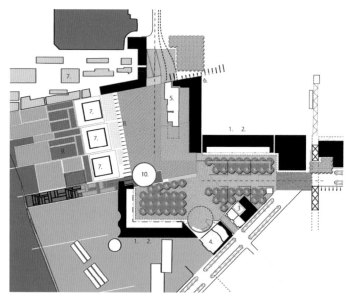

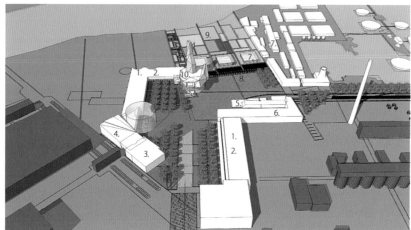

11 重要节点平面及意向 Highlights and Plan Concept

Civic Centre
1. Street Level Commercial
2. Office/hotel above street
3. Welcome/orientation Centre
4. Auditorium
5. Station
6. Parking/administrative
7. Chinese exhibition/pavilions
8. Re-purposed steel lattice screening
9. Reed run-off filtration beds
10. Digital Drum Tower/information node

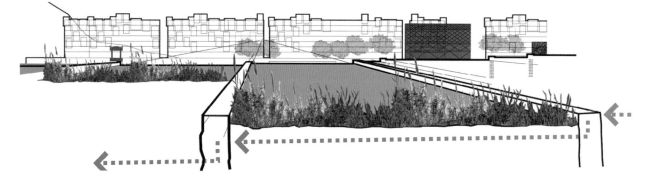

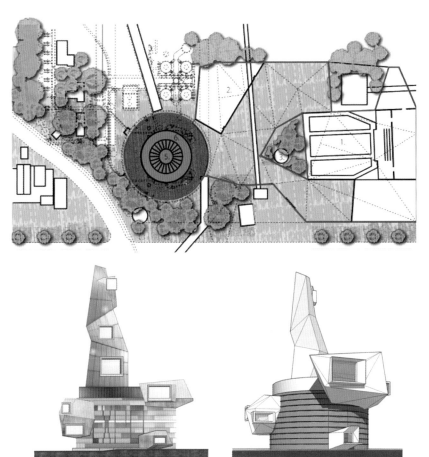

工业博览会集中在较大的湖周边，和南边的中国馆、国际馆以及北边的新的研究中心一起。大部分展馆将采用方便安装和拆卸的钢结构建成。

The Industrial Exposition is focused around the larger lake, with Chinese pavilions to the south, international pavilions within the lake itself, and a new research centre made up of research institutions and corporations to the north. Many pavilions are to be constructed of repurposed steel in a way that will make them easily adaptable and dismountable.

Furnace No. 4
1. Auditorium
2. Turf Roof
3. Central Courtyard
4. Chinese Industrial Archive No.1
5. Responsive Digital Archive
6. Furnace No.4

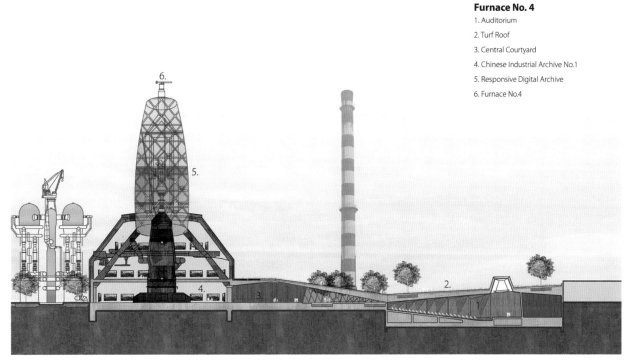

12 4号炉 Furnace No.4

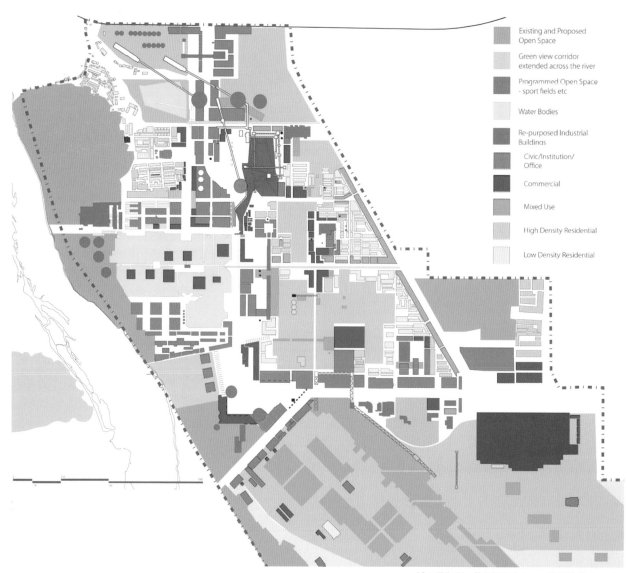

13 二期规划：技能博览会 Layer 2 :Skills Embassies

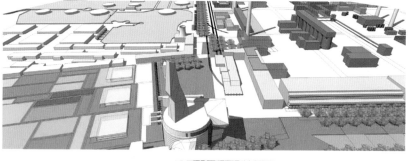

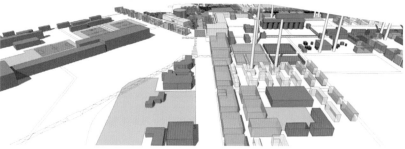

与世博会有关的功能和场馆成为未来的项目和土木结构的骨架，从而巩固和扩大场地上的展会功能。

Skills Embassies
The functions and pavilions associated with the Exposition become the skeleton of the future program and civic structure on the site. The new layer thus consolidates and extends the functions of the Exposition on the site.

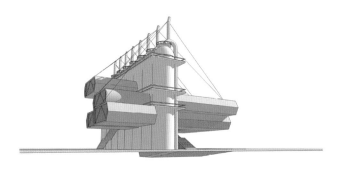

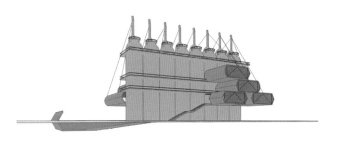

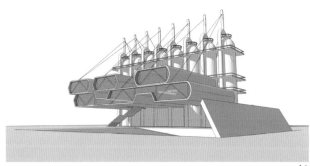

14 旧建筑改造 Reconstruction of the Old Building
15 透视图 Perspective

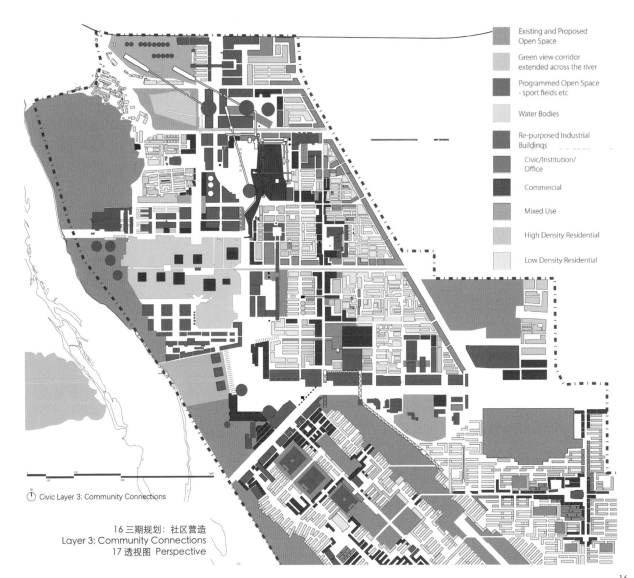

16 三期规划：社区营造
Layer 3: Community Connections
17 透视图 Perspective

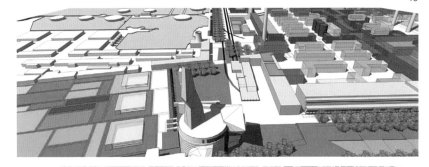

社区联系
这一阶段，商业、文化、商务功能的吸引性将会变强。

Community Connections
This layering sees a more liberal response to market pressures to the north-east of the site, given that the public and civic structure has already been underpinned here through previous layering. This phase sees the attraction of more commercial, institutional and business functions to the site as a result of the development of an industrial skills and innovation district here, with an associated community who support these new processes. The southern section of the site is developed during this layering as a community-centered quarter.

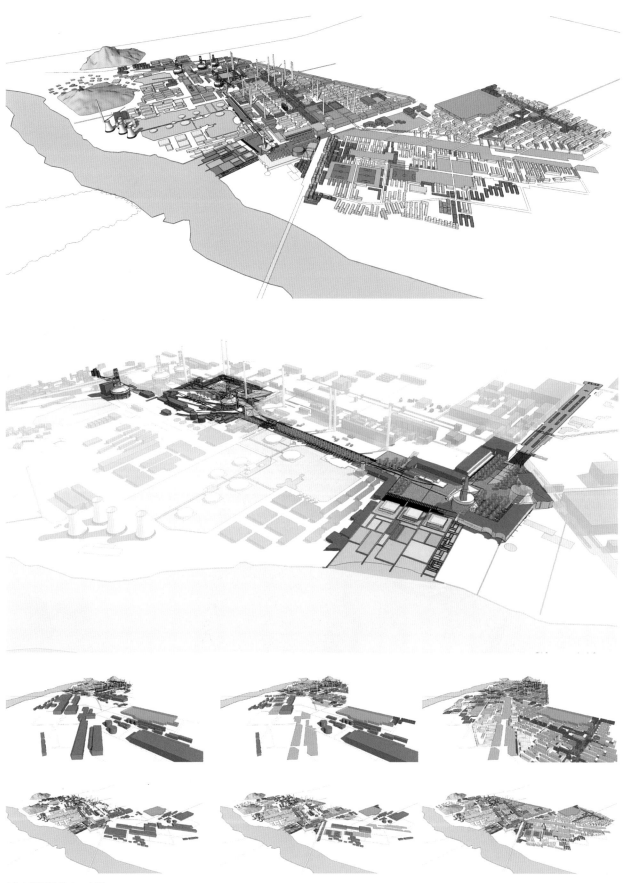

18 未来愿景 Future Vision

绿色首钢：全球可持续发展研究中心
GREENING THE FACTORY: A GLOBAL SUSTAINABILITY INSTITUTE IN BEIJING

Sonam Gayleg, He Zhong Yu, Jesse Hunting, Sarah Neilson, Sun Penghui, Jia Zheng

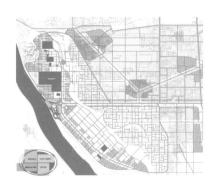

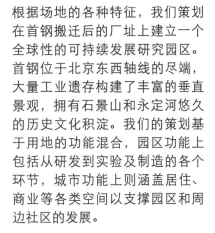

根据场地的各种特征，我们策划在首钢搬迁后的厂址上建立一个全球性的可持续发展研究园区。首钢位于北京东西轴线的尽端，大量工业遗存构建了丰富的垂直景观，拥有石景山和永定河悠久的历史文化积淀。我们的策划基于用地的功能混合，园区功能上包括从研发到实验及制造的各个环节，城市功能上则涵盖居住、商业等各类空间以支撑园区和周边社区的发展。

We propose the development of a Global Sustainability Research Institute at the Shougang Factory Site. Our plan responds to the many features at the Shougang site—its location on Beijing's east/west axis, the dramatic vertical landscape of manufacturing buildings, the height of Shijingshan Mountain and the presence of the historic Yong Ding River. We propose a rich mix of uses from research and development to testing and manufacturing, and an urban mix of housing and commercial facilities to support the Institute and its neighbors.

Total number of housing units: 18540
居住单元总数: 18540

Area east of green belt: 7800 units
Area west of green belt: 3540
Southern half of site: 7200

绿带以东: 7800套
绿带以西: 3540套
场地南段: 7200套

Total population: 51025
设计总人口: 51025

Area east of green belt: 20625 persons
Area west of green belt: 9750
Southern half of site: 20650

绿带以东: 20625人
绿带以西: 9750人
场地南段: 20650人

Commercial: 商业

Convention Center: 23690 square meters
Manufacturing: 371612 square meters
Commercial/office/retail: 1100000 square meters

会议中心: 23690平方米
制造业: 371612平方米
商业/办公/零售: 1100000平方米

1 总平面图 Master Plan
2 立剖面图 East-West Section

1

2

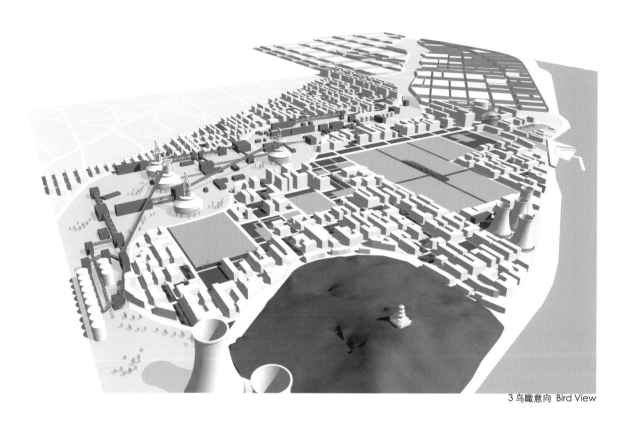

3 鸟瞰意向 Bird View

这一研究中心将使中国在能源效率、资源更新、水资源保护利用和交通系统创新等方面与国际接轨。研究机构、住宅、商业的混合以及对工业遗存的重新利用，会为人们提供一种全新的教育体验。我们策划若干国内外高等院校在此建立分校区，从事可持续发展的研究。

The Institute will position China at the center of an international movement toward energy efficiency, alternative energy, water conservation practices, and transportation innovations. Throughout the Institute, housing and commercial uses mix with historic structures adapted to new and sustainable uses, creating an array of educational experiences for visitors. A campus area features satellite facilities from leading universities in China and abroad.

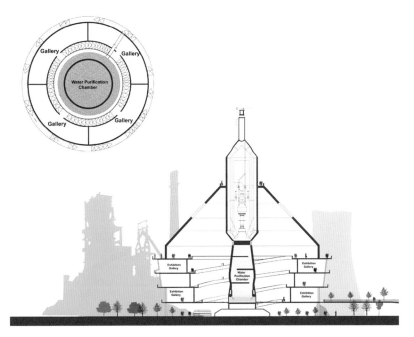

位于绿带中心的四号高炉被赋予游客中心的新功能，同时兼具展览、观景功能。盘旋而上的步行道在展示工业历史的同时引导游客步入高炉的顶点，这里为观看石景山、研究机构和绿带提供了全方位的视角。绿带中同时设计了各类研究成果展示的空间。

Furnace No.4, at the center of the green belt, is re-imagined as a dynamic Visitor Center, Exhibition Hall, and viewpoint. Spiraling walkways carry the visitor past exhibitions to the top of the furnace structure, to see panoramic views of the mountain, Institute and green area. Exhibition pavilions from Institute members, showcasing sustainability practices, are featured throughout the green belt.

4 重要节点图 Highlights

北京的东西轴线延伸穿过首钢的大门，终止于一个巨大的煤气罐前。这个曾经重要的工业设施现在被改造为新的地标，也是我们规划的交通枢纽。罐底与北京地铁相接。

Beijing's main east/west axis road leads the visitor to the historic Shougang entrance and beneath an enormous former gas tank, now thelandmark of the site's major transportation hub.

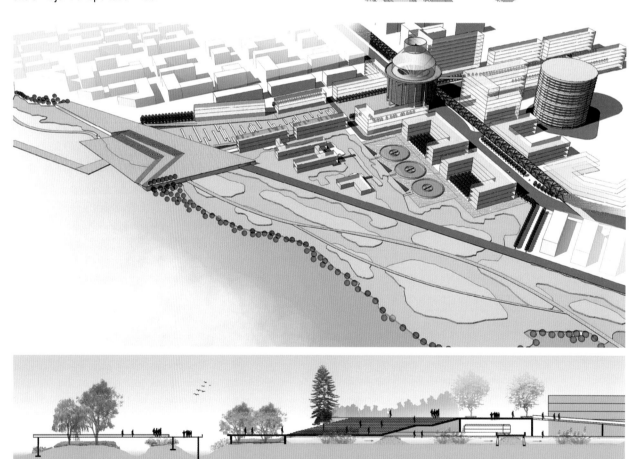

5 重要节点平面及意向 Highlights and Plan Concept

6 室内透视图 Interior Perspective
7 剖面图 Section

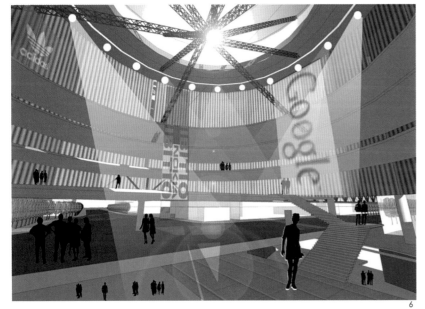

在罐顶人们则可以观赏永定河和新首钢及北京西部的城市空间。同时，我们也希望这里是外部私有交通工具和场地内部公共交通工具的转换点。利用首钢原有铁路系统的有轨电车贯穿整个场地，并在此汇聚，分布于地段各处的站点使人们可以在十分钟的步行时间内到达各处。

Visitors can enjoy site and riverfront views from atop the gas tank. Transportation throughout the site emphasizes alternatives to private automobiles and includes anchor points at the two Beijing subway stops on the site. A site-wide tram system sited connects all areas of the site to a tram stop within a ten-minute walk.

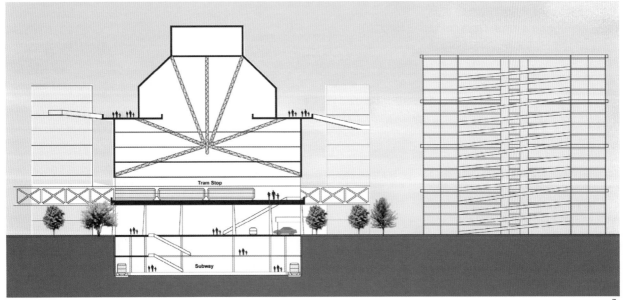

都市农田
URBAN AGRICULTURE:
A VERTICALLY-INTEGRATED APPROACH TO SHOUGANG'S SOCIAL, ECONOMIC AND PHYSICAL REVITALIZATION

Sandra Frem, Deborah H. Morris, Pamela Ritchot, Colin Zhao, Sara Zeng

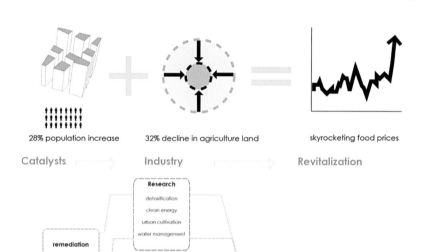

以棕地改造和城市农业为触媒，已设立研究中心。以城市农业生产和教育为手段，带动地区的经济复兴、社会复兴和空间复兴。

Shougang is an opportunity to form a global center for research surrounding brownfield remediation and urban agriculture. Rather than view shougang's broad-reaching pollution and industrial infrastructure as a constraint, this industrial heritage can be a catalyst for social, economic, and physical revitalization. Weaving agriculture in a hyperdense urban fabric generates a vertically integrated mixed-use environment where agricultural production and brownfield remediation intertwine with research and commerce surrounding food systems and modern ecology, educational institutions, agricultural tourism, living units and multi-modal transit.
This proposal responds to the paradoxical land use demands of explosive growth by resolving how to feed, house and employ the urban population.

1 场地分析图 Site Analysis

2 地段平面图 Masterplan

3 地段平面图 Illustrative Plan

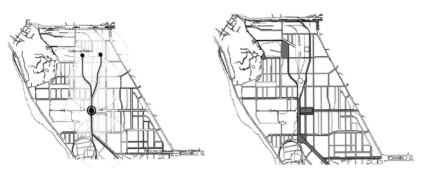

高密度中心区
vertical spine
required remediation:
expanded phytodegredation +extraction
remediation crops:
mustard, alfalfa, sunflowers, bamboo
food crops:
soybeans, potatoes, cabbage and green

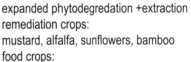

食品交易展区
food fair and convention center
required remediation:
expanded phytodegredation +extraction
remediation crops:
indian mustard,alfalfa,sorghum,barley, and rye
food crops:
soybeans,potatoes,cabbage,greens, fruits
soybeans, potatoes, cabbage and green

文化娱乐农业活动中心
agriculture tourism and recreation
required remediation:
phytodegredation +extraction
remediation crops:
indian mustard,alfalfa, sunflowers, bamboo
food crops:
soybeans, potatoes, cabbage and greens

4 规划结构 Development Scheme

中心车站未来将成为进入新首钢地区的门户，它位于高密度哑铃型中心区的中央，出入口直接面向群明湖。

中心车站既作为过境火车的到达车站，也连接一号线延长的地铁，并且作为内部有轨公交车发车站，它将成为地段最重要的交通综合体。设计中考虑不同流线交通的分流：地下——地铁／火车，地下一层——停车；地上——行人／出租车。

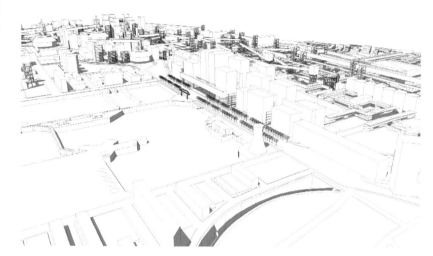

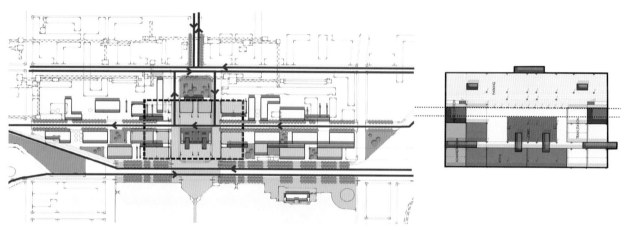

5 重点节点平面 Detail Plan

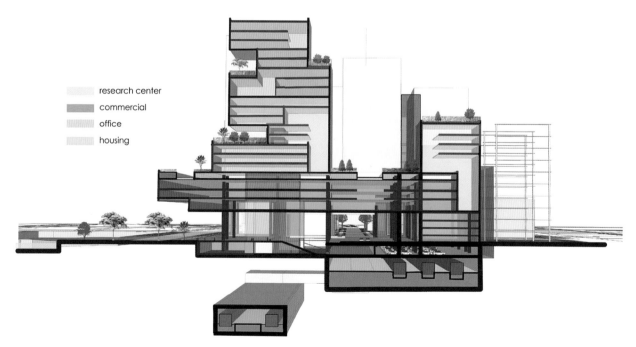

6 重点节点剖视图 Detail Section

024

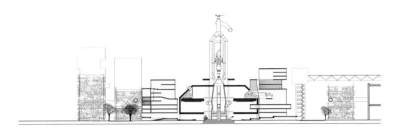

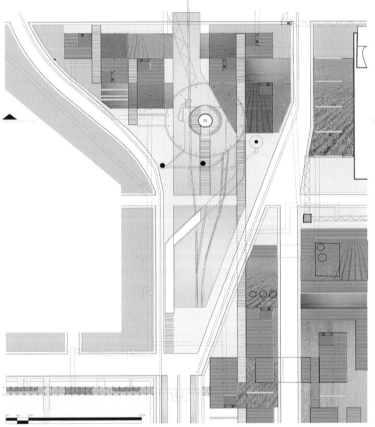

四号高炉是地段的重要节点。它是首钢复兴概念"城市农业"的展示窗口。未来将改造为棕地改造和都市农业的技术展览和信息交流中心。

Creating a hyperdense, vetically-productive spine within shougang, makes it both economically feasibel and beneficial to have large plots of land open for natural phytoremediation. Doing so avoids the high capital costs of soil washing and extraction while creating a valuable research asset: the remediation itself.

7 重点节点：食品交易区 Detail: Food Fair

2010

济南　JINAN
居住区　NEIGHBORHOOD
清洁能源城市

高铁建设从区域格局上提升了城市的机动性与可达性。济南高铁新城片区位于济南中心城区的西侧，是一个经过系统规划与城市设计的全新片区。在黄河与南部山区所限定的东西轴向发展的空间形态中，成为济南城市空间的西侧端头。在这样一个全新的发展背景下，如何考虑城市的可持续发展、生活与工作的融合、低碳的空间形态，是本次规划设计研讨的重点。在原有规划的基础上，选择数个地段，分别从生态环境、自然资源、用地结构、交通方式等方面，从城市设计的角度，对上述诸多方面进行深入研讨，并利用清华—MIT 在"清洁能源城市"研究中的模型，对碳足迹进行计算和追踪，优化规划设计方案。

High speed rail construction enhanced the city's mobility and accessibility in regional level. Jinan high-speed rail new town is located in the west side of the city center, which is a well-planned new town going through systematical urban planning and design process. The Yellow River and the southern mountain enclosing an east-west axial space, where Jinan located in the west end of the axis. Under such a new development background, how to achieve the sustainable development of the city, how to integrate the life and work of people, and how to formulate low carbon strategies, are the key issues of this urban design project.On the basis of the original planning, several sites are chosen to show intensive study in ecosystem, natural resources, urban structure, transportation and other aspects from the perspective of urban design. The urban design project will be optimized by the model using in the research of "Tsinghua-MIT Clean Energy City" and the calculation applied to carbon footprint.

社区生活，绿色城市
COMMUNITY LIFE, GREEN CITY

Timothy Bates, Matthew Bindner, Alissa Chastain, Erik Scanlon, Wang Chao, Zhai Bingzhe

该地段毗邻中国济南市的西部新城，规划设计中将会把高层商业和低密度住宅用地结合起来。设计的模式与该地区的能源战略将会创建一个具有优美形态和混合使用的环境，从而满足高密度发展的需求，并创造一个非常适宜居住的街道氛围。现有的河道是整个社区的轴线，将社区串联起来，并在设计的过程中成为社区生活的焦点。

This new neighborhood district is located adjacent to a planned downtown in west Jinan, China, and will connect high-rise business and low-rise residential areas. The form and energy strategy for the district will create a fine-grained, mixed-use environment that allows for high-rise development and provides a vibrant street life. An existing canal will be relocated to form the central spine through the neighborhood, becoming a focus for community life.

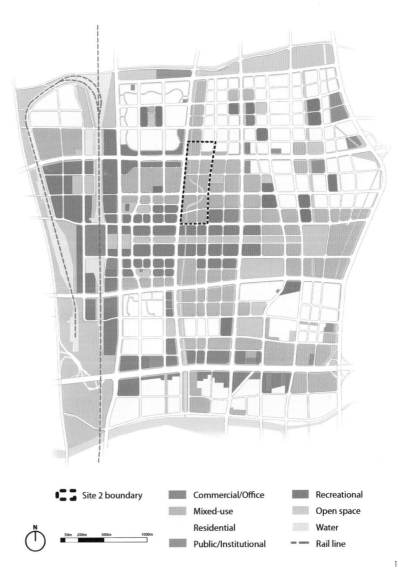

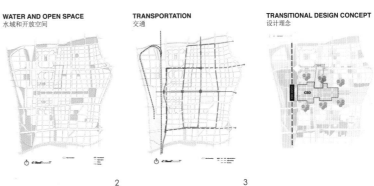

1 地段用地性质图 Land Use
2 水域和开放空间 Water and Open Space
3 交通 Transportation
4 设计理念 Design Concept

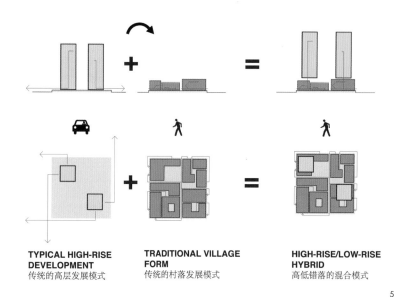

TYPICAL HIGH-RISE DEVELOPMENT	TRADITIONAL VILLAGE FORM	HIGH-RISE/LOW-RISE HYBRID
传统的高层发展模式	传统的村落发展模式	高低错落的混合模式

5

清洁能源城市的手段与方法
1. 采用混合的、适合发展规模的能源方式，同时使用积极的和消极的策略，以减少能源的使用和提高能源使用的效率
2. 建造一个适应于物质和社会条件不断变化情形的建成环境，用来减少重建、改建的需求
3. 提供一个完整的社区用以满足居民较大程度上的日常需求，这样可以减少机动车以及能耗交通的使用
4. 获取可再生资源，使用地热能源的利用系统

设计准则
1. 规模以及密度的多样性　在减少电梯使用的同时，凸显不同的层级，提供空间感以及满足容积率的要求
2. 适宜步行的完整社区　在居住环境周围提供日常生活的必需品和服务以及愉悦的环境，以此来减少非步行的交通需求
3. 将水体作为组织的元素　通过设计和形态的组织在基地内凸显水体

DESIGN PRINCIPLES
•**Multiple scales and densities**, to convey hierarchy, provide a sense of space, and meet FAR needs, while minimizing elevator use
•**Walkable neighborhoods**, to provide amenities and daily needs goods and services in close proximity to residences, thereby reducing the need for transportation trips
•**Water as an organizing feature for the neighborhood**, using the site to highlight water issues through form and programming
•**Adaptability**, to extend the life and utility of the built environment for residents and users becoming a focus for community life

CLEAN ENERGY CITY APPROACH
•**Employ an integrated, development-scale energy approach**, using both active and passive strategies, to reduce energy use and increase efficiency
•**Create a built environment that can adapt to changing physical and social conditions**, reducing the need to rebuild or move away
•**Provide "complete" neighborhoods that serve a wide variety of daily needs for residents**, reducing the need for vehicle/ transportation trips
•**Capture renewable energy**, using geothermal heat exchange

能源策略的结果
结合的社区形态看上去可以达到我们预设的目标——高层形态与低层形态的结合产生了高度混合利用的社区，而其高度的可达性则可以达到节能的目的，提供足够的容积率以及满足日照的要求，所有的一切都是可以实现的设计，并且能够在城市的不同尺度和范围下复制利用。

ENERGY STRATEGY RESULT
•**Combining neighborhood typologies appears to achieve intended result** – Combining high-rise and low-rise neighborhood typologies yields a highly mixed-use neighborhood with high accessibility that can save energy, provide sufficient FAR, and meet sunlight requirements, all within a realistic design that can be replicated throughout a city at different scales.

5 区位分析　Location Analysis
6 发展模式　Developing Pattern
7 多样化体验　Diverse Experience

发展模式的改变

济南市位于中国首都北京的南部，距离北京大约500公里。济南是山东省的省会城市，并以"泉城"而闻名。位于济南西部新城的高速铁路将会连接北京和上海，同时，一个新的包括商务、财经、艺术文化以及居住等功能的城镇中心将会围绕高铁车站营建。

SHIFTING DEVELOPMENT PATTERNS

Located approximately 500 kilometers south of China's capital, the City of Jinan is the capital of Shandong Province and is known as the "city of springs." A new high-speed rail stop will connect the city to Beijing and Shanghai, and a new town center will be constructed around the station, providing an opportunity to create a more energy-efficient district and shift typical development patterns.

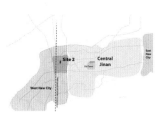

社区形态

传统的低层发展模式能够为人们营建出高质量的生活起居环境，但是却不能满足中国城市土地紧缺的密度要求。相反，典型的高层发展模式能够提高土地的使用强度，但是却不能构建出好的城市肌理。而我们会把这两种模式的优点结合起来设计这个新的地段。

COMBINING NEIGHBORHOOD FORMS

Traditional low-rise development can create high-quality places to live and workin, but cannot provide the density demanded in modern Chinese cities. Conversely, typical high-rise building typologies provide density, but do not create excellent urban fabric. The new site will combine both types to capture on the benefits of each.

为居民提供多样化的城市生活体验

通过串接一系列的活动以及良好的生活环境，社区将会为居民和使用者提供积极的生活体验，而不是将这些体验置于城市其他的地方。

PROVIDING A DIVERSE URBAN EXPERIENCE

The neighborhood will provide residents and users a positive experience by clustering a range of activities and amenities, rather than placing them in discrete areas of the city.

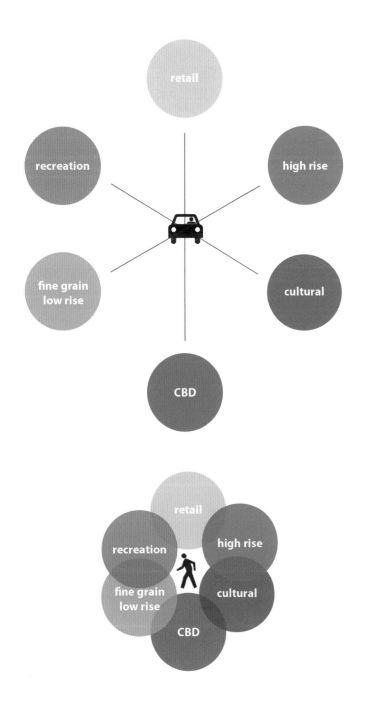

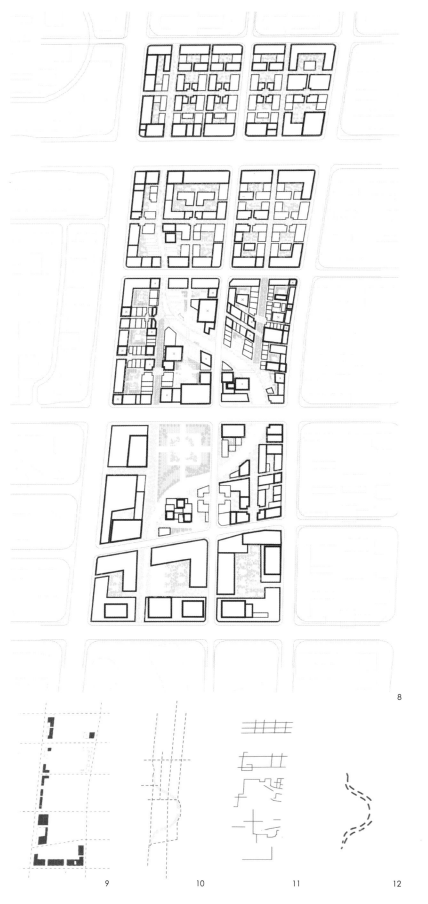

整体规划围绕着内部的水域将社区分成若干活动的片区。在水域附近组织了环形交通、土地利用和建筑的大致形态。整个社区分成明显的三个区域，这些区域显示了从高密度活动区到低密度活动区的转换与改变。所有的片区都反映了庭院理念以及一体化建筑的理念。

The overall plan organizes the neighborhood into zones of activity around the river form. Circulation, land use, and massing are all oriented around the water. The neighborhood has three distinct areas that demonstrate the transition from high density and activity to lower density and quieter places. All areas reflect the courtyard and integrated building concept.

对于北部的居住社区，还包含一些社区的公共服务设施以及购物场所。社区的中心采用多重复合土地利用的手段，并围绕河流组织了步行活动。南部片区主要是一些商业用地，还包括一些关键的文化娱乐设施与机构。

To the north, the neighborhood is primarily a residential district, with supporting community services and shopping. The center of the neighborhood is characterized by highly mixed land uses and pedestrian activity around the river. The southern portion is predominantly commercial and contains several key cultural institutions.

8 总平面图 Floor Plan
9 机动车与停车 Parking
10 快速自行车 / 步行 Fast Speed Bicycle/Walking Routes
11 中速自行车 / 步行 Medium Speed Bicycle/Walking Routes
12 慢速自行车 / 步行 Low Speed Bicycle/Walking Routes
13 总体土地利用 Overall Land Use
14 建筑体量与电梯 Building Blocks and Elevators

关键理念

1. 完善的社区，在近距离享有日常的购物需求、服务设施以及居民需要的优美的环境。
2. 高层建筑加上具有优美肌理的低层建筑。这些建筑会为整个社区营建出一个人性尺度的街道肌理，与较高密度的紧凑发展模式并不矛盾。
3. 有边界的街区庭院，用以确定边界，创建微气候，为居民提供开放空间。
4. 一体化建筑的理念。通过串联低层建筑与高层建筑"风塔"将空气加热设备和制冷设备连接起来，提供自然通风。
5. 地热泵的使用。在一些有热井和冷井的庭院使用利用地热能量的设备。
6. 利用艺术与文化设施以及公共空间将商务区与居住区之间的社区连接起来。
7. 河道营建出为社区提供特殊活动的中心地，而且进一步连接了各种娱乐休闲活动的三个片区。
8. 有梯度的密度模式为居住单元提供了阳光，但是也满足了高密度发展模式的要求。

KEY CONCEPTS

- **"Complete" neighborhood**, containing the daily needs shopping, services, and amenities that residents need within close distance
- **High-rise + fine grain low-rise**, resulting in a neighborhood that has a human-scale street fabric, but also allows for higher-intensity development
- **Perimeter block courtyards**, to define edges, create microclimates, provide open space for residents
- **Integrated building concept**, creating a connected heating and cooling system that combines walk-up housing with "wind tower" building to provide natural ventilation

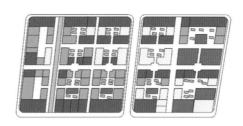

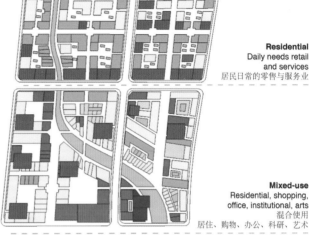

Residential
Daily needs retail and services
居民日常的零售与服务业

Mixed-use
Residential, shopping, office, institutional, arts
居住、购物、办公、科研、艺术

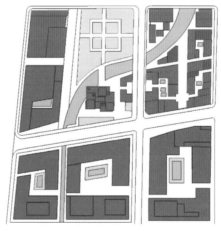

Commercial
Arts and culture institutions, comparison shopping, office
商业
艺术、文化机构，商业、办公的比较

13

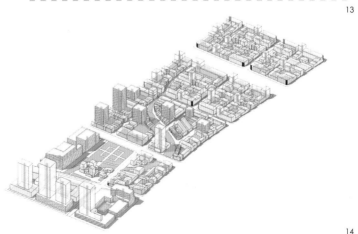

14

RIVER AS THE HEART OF THE NEIGHBORHOOD
作为片区核心的河流

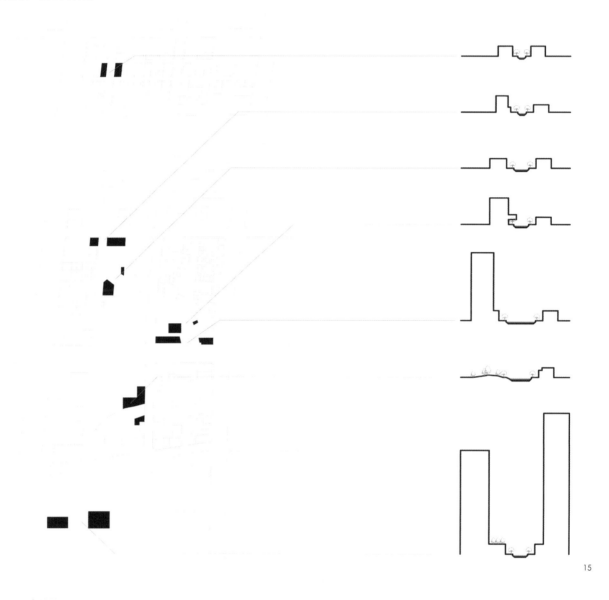

- **Geothermal heat pumps**, with warm and cold wells located in alternating courtyards
- **Transition from business district to residential neighborhood**, using arts and culture institutions and open space as elements to connect these areas
- **River creates zone of special activity and a central place for the neighborhood**, further uniting the three zones of the district and focusing recreation and leisure activities
- **Density gradient provides light to residential units**, but allows higher density development.

15 片区核心河流 River as the Heart of the Neighborhood
16 文化中心 Cultural Center
17 街坊规划 Neighborhood Planning
18 垂直开放空间系统 Vertical Open Space System
19 垂直系统设计 Vertical System Design

17

街坊规划将场地概念融入中心混合使用区的一个街坊中。这个街坊主导了许多不同规模的空间以及不同类型的活动，而河流已经作为所有活动的焦点。

The cluster plan translates the site concept to the scale of one large block located in the heart of the mixed-use district. The cluster is host to many layers of activities and scales of space between. The river continues to be the primary focus of activity.

从南至北的步行街道和稍小规模的东西向人行路线将街坊分隔得更具人性尺度，而且提供了高度可达性的肌理结构。通过特色的植被和营建微气候的水池，这些线路将半私人的庭院连接起来。

Pedestrian streets running north to south and smaller east-west pedestrian routes break the block into human scale pieces and provide a highly accessible fabric. These routes connect semi-private courtyards, which feature vegetation and pools to create microclimates.

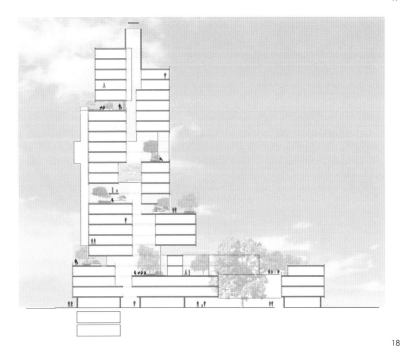

18

VERTICAL MIXED-USE
垂直方向上的土地混合利用

CONNECTED VENTILATION SYSTEM
连接起来的通风系统

PUBLIC AND PRIVATE OPEN SPACE
公共开放空间以及私人开放空间

19

035

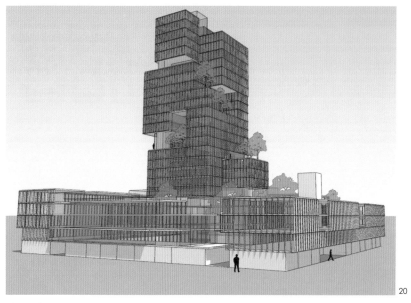

关键理念

1. 通过河流来定义街坊的组织形式，打破了方格网的形式而且为街坊带来了更多的阳光。
2. 封闭的街坊庭院是位于河流之后的另外一个重要的组织要素，庭院不仅仅定义了公共空间和私人空间，同时也为地热能源利用系统以及共享的通风系统提供物质场地环境。
3. 一体化建筑的理念将各种类型的建筑的通风系统连接起来，以提供最自然化的通风条件。
4. 南北向的步行街道是购物的核心场所，这里有大大小小的店铺，街道的报刊亭以及各种街道活动等。
5. 河流区域是一个休闲区域。这里有餐饮活动、休闲活动以及小商店和美术馆。
6. 遮蔽的策略包括了对高层建筑的精心选位和临时性的遮蔽物，以防止对居住区过多的遮挡；而且为庭院提供夏日的阴凉。
7. 各个层级的空间包括公共街道，公共人行道路，半私人庭院以及私人的空中室内花园。
8. 街坊内部对于步行和自行车交通开放，可达性很强。

20 低层建筑位于边缘，高层建筑位于后方 Low-rise Buildings on the Edge, High Rise Buildings Behind
21 河岸鸟瞰 Riverbank Birdview
22 河岸边缘透视 Riverbank Perspective
23 步行街透视 Walking Street Perspective

KEY CONCEPTS

- **Primary organization of cluster defined by river**, breaking the grid pattern and opening the block for more sunlight
- **Perimeter block courtyards** are the secondary organizing element after the river, not only defining public and more private spaces, but also creating space for geothermal pumps and housing a shared heating and cooling system.
- **Integrated building concept**, creating a connected heating and cooling system that combines walk-up housing with "wind tower" building to provide natural ventilation
- **North-south pedestrian street is center for shopping**, with small and large storefronts, street kiosks, and street activity (such as performers)
- **River is a more relaxed zone**, for dining, relaxing, small shops, and galleries
- **Strategic shading** includes careful location of high-rise buildings to avoid excessive shade on residential units but to provide summer shade for courtyards, as well as temporary shades/awnings
- **Hierarchy of space** includes public streets, public pedestrian ways, semi-private courtyards, and private elevated winter gardens
- **Interior of cluster accessible via pedestrian and bicycle traffic only**

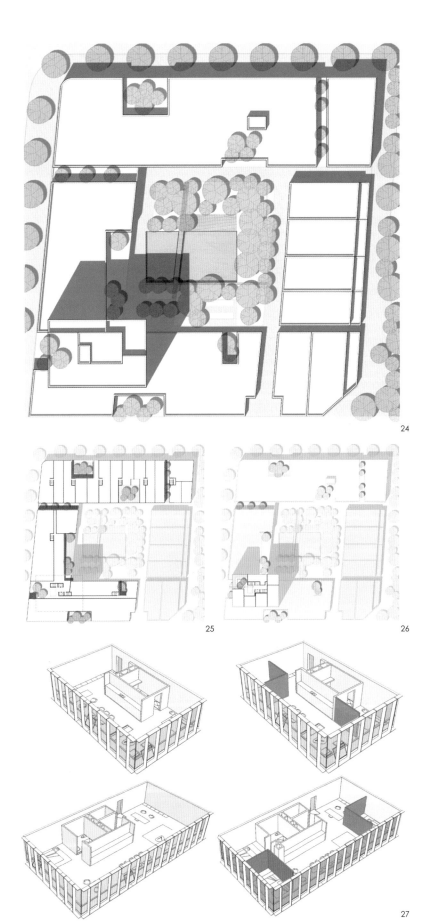

24 庭院平面 Courtyard Plan
25 混合使用四层建筑平面
4th Floor Plan of Mixed-use Building
26 典型高层建筑平面
Typical High-rise Building Plan
27 单元可变性 Unit Alterability

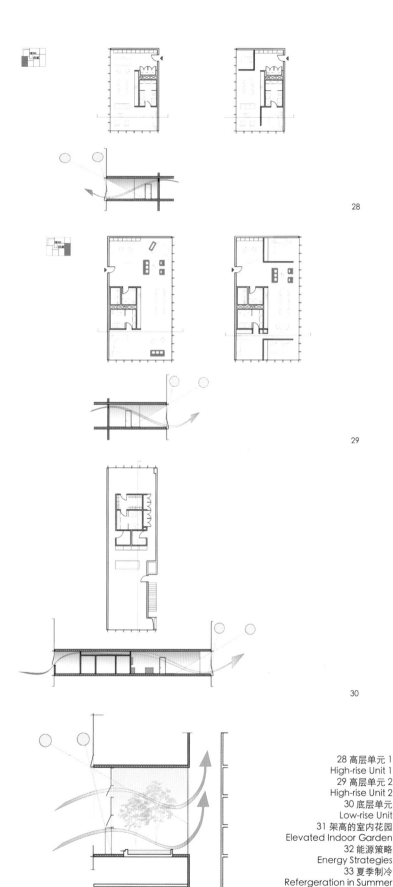

28 高层单元 1
High-rise Unit 1
29 高层单元 2
High-rise Unit 2
30 底层单元
Low-rise Unit
31 架高的室内花园
Elevated Indoor Garden
32 能源策略
Energy Strategies
33 夏季制冷
Refergeration in Summer
34 冬季取暖
Heating in Winter

建筑和单元策略

1. 典型的建筑模式结合了高潮的塔楼同时兼容了步行的庭院建筑模式
2. 综合的建筑热量交换系统：通过集中通风的中庭将单元连接起来，以消除季节性的对空气加热和制冷设备的依赖
3. 建筑楼板是混凝土的核心板，能够使得单元间改善或者加热的空气交换成为可能
4. 单元开放的楼层平面可以跟随住户的需求而改变，通过这样的方式还可以营建分散的温度区域
5. 开放的楼层平面通过天井连接起来，以获得最大程度的水平通风
6. 每个单元都能保证够获得要求的日照
7. 在不同的楼层设置室内花园和绿化的屋顶
8. 庭院用以营建微环境来改善低层的建筑单元环境和片区的通风系统
9. 集中水、电、暖等管线来提高使用效率，增加可用的空间，提高单元空间的可变性

BUILDING AND UNIT STRATEGY

- Typical building combines a high-rise tower with walk-up courtyard building
- Integrated building heat exchange system: Units connected to central ventilation atrium to expel either hot or cold air depending on season
- Building floorplates are constructed of concrete, core slabs, which allow for unit-to-unit exchange of conditioned or heated air
- Units have open floorplans to allow for adaptability as resident needs change, as well as a way to divide space to create discrete temperature zones
- Open floorplans work in conjunction with central atrium to allow for maximum cross-ventilation
- Each unit receives minimum required sunlight
- Building integrates green roofs and winter gardens at different levels
- Courtyard creates microclimate to benefit lower units and ventilation system
- Plumbing, mechanical, and electrical cores are centralized to maximize material efficiency, volume of usable space, and adaptability of units

清洁能源城市策略

1. 综合的建筑热量交换系统——适宜发展规模的解决方式：在庭院周边设置有空气调节功能的高层建筑，于冬夏两季在建筑单元间能够采用技术手段来引导空气流动循环

2. 微气候：庭院、水池以及提供被动温度调节的绿化系统

3. 策略性遮蔽：建筑布置、建筑细节以及遮挡面

4. 可改变的居住单元：可改变的用途，建筑单元，楼层平面等能够最大限度地利用空间以及提供通风环境

5. 有限的电梯使用：步行建筑占主导，选择性地布置高层建筑的使用电梯

6. 太阳能运用：通过太阳能量的获取为居住者提供阳光需求

7. 减少小汽车使用：使得片区具有高度可达性，为步行道路营建宜人环境，有限的地下停车以减少汽车行驶能量获取时间和行驶距离

CLEAN ENERGY CITY STRATEGIES

- **Integrated building heat exchange system** –"Development-scale" solution: perimeter courtyard with high-rise "cooling tower" with mechanical systems to direct airflow and circulation throughout the block in both hot and cold seasons.
- **Microclimates** – courtyards, pools, and vegetation to provide passive cooling
- **Strategic shading** – building placement, architectural details, and screens
- **Adaptable living units** – adaptable uses, units, floorplans provide ventilation and extend the utility of units
- **Limited elevator use** – predominantly a walk-up district, with selectively placed high-rise
- **Solar orientation** – solar heat gain, meet residential sunlight requirement
- **Reduce car usage** – highly accessible, high amenity pedestrian streets, limited underground parking to reduce vehicle trips and trip length

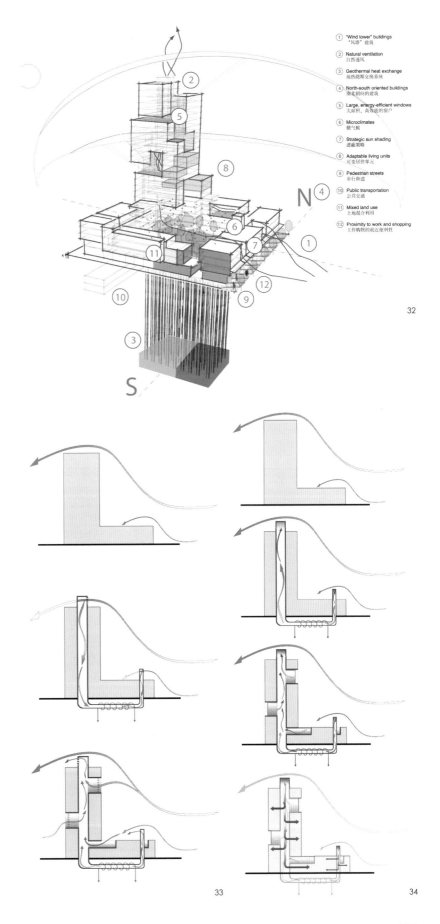

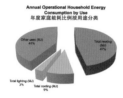

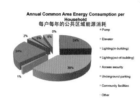

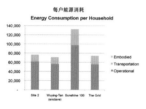

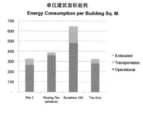

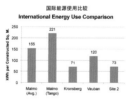

能源形式验证

1. 用以温度调节的能源是能源使用最大的一部分，超过 50% 的能源消耗来自于空气制冷和加热过程。
2. 地段 2 中每个家庭的平均能源消耗会比济南许多地段水平略微高一点，但是每一建筑区域在节能方面效果更好，过去的形式不能充分地展示出不同功能用地的比例和构成。
3. 过去的形式不能充分地获取与规模适宜的空气温度调节系统的效果，这样的系统不单单是一般意义上的能量交换和局限性的空气调节。
4. 能量使用结果和国际上的标准相一致，但是需要更多的调查和研究，目前还不清楚这样的模式是不是与原始输入的数据相一致，除此之外，这里对社会经济学的不同并没有进行讨论。

混合的社区形式是更加有效节能的形式并且能够让这一地区更加适于居住

高层模式与低层模式混合运用下的新模式在能源方面比一般的典型高层模式更加出色，和济南的老城区的能源消耗相持平。这种混合运用产生了多样化的、混合土地利用的社区，为居民们提供了每天必需的宜人环境，同时满足了日照和容积率的要求。除此之外，这样的模式强调了步行交通，降低了其他交通能耗。这种模式的现实性很强，可以在同一城市的不同地段、不同规模尺度上得以复制和利用。

ENERGY PRO FORMA RESULTS

Operational energy is largest part of energy use – Over 50% of consumption is used for heating and cooling.

Energy use per household in Site 2 is slightly higher than many Jinan typologies, but performance per constructed building area is better.

– The pro forma does not adequately address the proportion of different land uses.

Pro forma may not be adequately capturing effect of development-scale building heating and cooling – This is a system that goes beyond normal heat exchange or district conditioning.

Energy use result compares favorably to international prototypes, but more research is needed – It is not clear whether the pro-forma result represents exactly the same inputs as the prototype data. Further, there is no control for socioeconomic differences.

HYBRID NEIGHBORHOOD FORM IS MORE ENERGY EFFICIENT AND MAKES MORE LIVABLE PLACES

Merging the high-rise and low-rise neighborhood typologies results in a neighborhood form that outperforms the energy consumption of "typical" highrise development and is equivalent to older, lower density areas of Jinan. This hybrid yields a diverse, mixed-use neighborhood that provides the daily amenities that residents need, meets sunlight and FAR requirements, and that emphasizes pedestrian and non-automobile travel. The form achieves all this within a realistic design that can be applied throughout a city at different scales.

35 能耗分析图
Energy Consumption Analysis
36 地段能耗比较
Site Energy Consumption Comparison
37 家庭能源消费表
Energy Consumption Per Household
38 单位建筑面积能源消费表
Energy Consumption Per Building Square Meter

Site 2 Comparison 地段 2 比较

	Site 2	Wuying-Tan (enclave)	Sunshine 100	The Grid
Population	16,242	16,100	19,000	11,700
Pop. Density (# 1000 ppl/sq km)	32	36	54	38
FAR	2.54	1.16	2.15	1.69
Building Coverage	0.40	0.28	0.14	0.31
Average Building Height (# of floors)	6.35	4.18	14.91	5.47
Land Use Mix	0.41	0.29	0.26	0.30
Car Ownership (vehicles/100 HH)	55	21	68	16

Energy Consumption per Household (MJ) 家庭能源消费 (MJ)

	Site 2	Wuying-Tan (enclave)	Sunshine 100	The Grid
Operational	61,442	56,362	97,645	55,785
Transportation	4,077	8,484	23,657	7,352
Embodied	11,311	6,979	11,055	11,435
Total	76,830	71,825	132,357	74,572

Energy Consumption per Building Square Meter (MJ) 单位建筑面积能源消费 (MJ)

	Site 2	Wuying-Tan (enclave)	Sunshine 100	The Grid
Operational	262	358	480	280
Transportation	17	17	116	19
Embodied	48	14	54	30
Total	327	389	650	329

活力社区
VIBRANT COMMUNITY

Sai He, Elizabeth Ramaccia, Sagarika Suri, Tony Vanky, John Xie, Aspasia Xypolia, Yang Yang

这是一个为济南新商务中心区所制定的一份原生态、高节能、高生活品质的计划书。这个设计的理念便是要将人与自然进行结合,针对现代社会以及一些可变动的社区因素使人们更加能够融入这个社区。

水、光、风,这些自然因素改变了我们在这一地区的设计理念。从社区整体,建筑组团到每个个体,犹如浑然天成。在这个变性的空间中,我们考虑了时间与距离的问题,并增加设立社区通道以方便大众。这个被我们称为"副无车区"的交通方式使人们即使在这里骑车或者步行,也会感到十分的舒适,安全。

This proposal for a new central business district for the city of Jinan, China is an alternative model to typical contemporary city planning in China and serves as a prototype for city construction that is energy-efficient and provides a high quality of life. The design acknowledges the direct correlation between the necessity to design with environmental systems and possibilities for vibrant community.

Water, sun, and wind systems shape the design at the regional, site, building cluster, and units scales. Flexible and temporal spaces adapt to changing needs across seasons and over time. Increased connectivity across the region is created through a secondary non-vehicular transport system for bicyclists and pedestrians that is both efficient and pleasant.

1 地区尺度 Regional Scale
2 多中心 Multi-center
3 交通连接 Traffic Connection

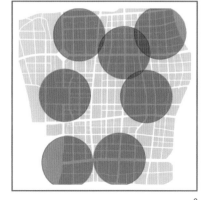

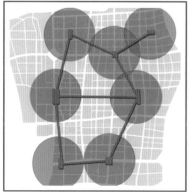

Energy strategies at various scales
多尺度能源策略

Proposed community centres 4

Solar
阳光

Proposed commercial areas 5

Water
水

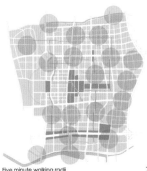
Proposed commercial centres 6

Wind-Ventilation
通风

Landscape
景观

Five minute walking radii 7

Flexibility
灵活性

Connectivity-Mobility
连通性

Water and green infrastructure circulation systems 8

我们利用人文以及自然生态打造出了一个可以对机动车做出限制的新的生活、工作方式。通过这个新的方式，可以有效地节约社区中心到每个交通站的时间。我们还建立了一套强有力的水上交通系统，水渠东西向跨越整个社区，连接了各个社区中小型的中心，这可以不用增加机动车道路而减小街区的面积。水渠交通给行人和自行车使用者提供了地标指示，并潜在地使人们感觉骑车、步行将会比开车更加节约时间。另外，水渠可以很好地收集雨水，并通过废水回收系统重新使用于周边地区的农业上。

The utilization of man-made and natural ecological systems on the regional scale supports new ways of living and working that discentivize the automobile yet increase time-effiency between neighborhood centers and transit stops. A dynamic system of canals running east-west across the region connects sites identified as centers to a central canal and breaks down block sizes without the addition of vehicular roads. The canal system, in tandem with a north-south running system of green paths, informs a secondary bicyclist and pedestrian transport system and acts as a way finding device. The potential for a scenario is created in which biking or walking is actually more time-efficient than driving. Additionally, the canals organize a rainwater collection and grey-water reclamation system to capture all water received and used on site and support new endeavors in urban agriculture.

4 建议社区中心 Community Center
5 建议商业空间 Commercial Area
6 建议商业中心 Commercial Center
7 五分钟通行圈 5-min circle
8 水系绿化系统 Green Space System
9 土地利用 Land Use
10 日照通风分析 Sunlight and Ventilation Analysis
11 地段尺度 Regional Scale

虽然区划是全部应用于商业使用，不过加入一些住宅计划会使这个地域更能够适合不同人群。商业建筑坐北朝南，坐落于整个地段的南部。夏季时分，东南风渗入建筑群。东西走向的住宅建筑将位于商务建筑的上方，这可以使日照最大化地渗入人们的住宅。在地段的北部，一些建筑群的布置阻止了冬天的北风。地段从北到南，建筑由高到低，以此来捕捉阳光。整个地段的容积率大约3.7左右。绿地将在多个楼层间出现。

Although originally zoned to be entirely commercial, the addition of residential programming is necessary to create a place that is inhabited throughout the day by various user groups. Staggered commercial buildings are oriented north-south in the southern portion of the site to allow southeastern summer winds to penetrate the site. East-west-oriented residential is situated atop commercial buildings to allow maximum daylight penetration. On the northern portion of the site, programs and placements are reversed to block northern winter winds. Buildings become gradually taller from south to north across the site again to capture sunlight. The site achieves an FAR of approximately 3.7. Green space begins to occur at multiple vertical levels.

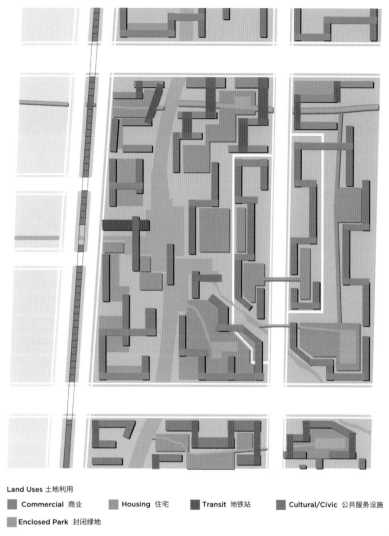

Land Uses 土地利用
- Commercial 商业
- Housing 住宅
- Transit 地铁站
- Cultural/Civic 公共服务设施
- Enclosed Park 封闭绿地

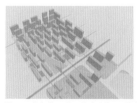  

Diagrams illustrating system of orientation and building heights for ideal solar exposure and site ventilation
为获得理想的太阳能照射和通风的建筑定位和高度图示

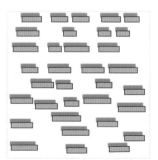 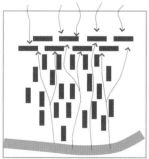 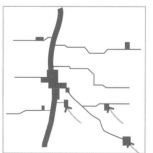 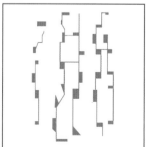

Strategic orientation of facades for sunlight
立面日照策略定位

strategic orientation of building masses for natural ventilation
建筑群自然通风策略定位

Pool of water on south east corner for cooling summer winds
用来冷却夏季风的东南角水池

Green paths for cycling and pedestrian movement
自行车和行人的绿色运动路径

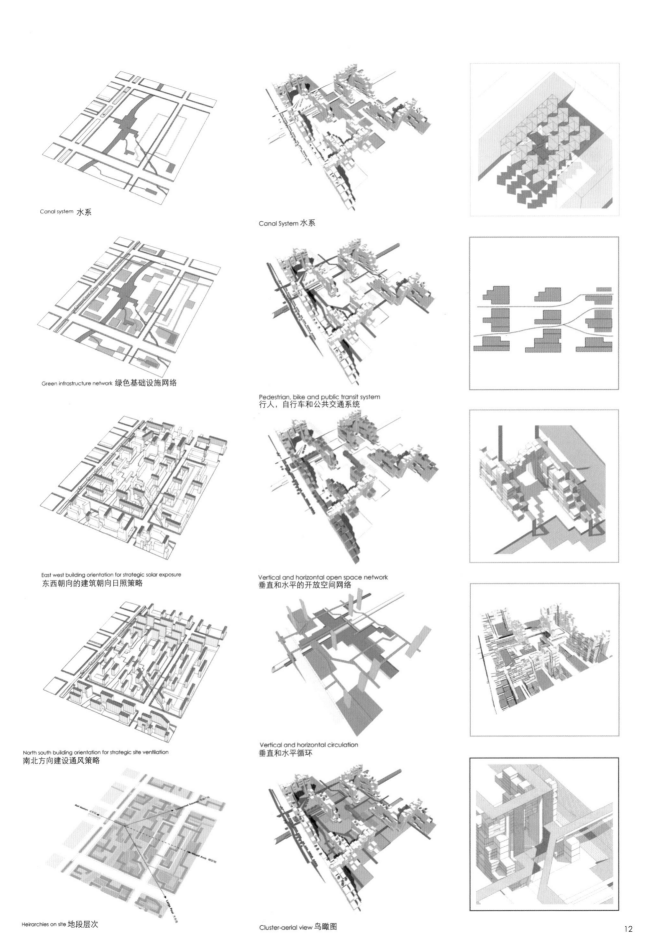

Canal system 水系

Canal System 水系

Green infrastructure network 绿色基础设施网络

Pedestrian, bike and public transit system
行人，自行车和公共交通系统

East west building orientation for strategic solar exposure
东西朝向的建筑朝向日照策略

Vertical and horizontal open space network
垂直和水平的开放空间网络

North south building orientation for strategic site ventilation
南北方向建设通风策略

Vertical and horizontal circulation
垂直和水平循环

Heirarchies on site 地段层次

Cluster-aerial view 鸟瞰图

组团尺度

这里定义的组团包括了特征空间地段以及独一无二空间的设定。组团包含了主要的地铁站和地域文化中心。这个组团是水渠与城市的分割区,不同于其他地方行人可以和水渠有一个良好的互动。

建筑中的穿孔和阶梯形构造可以增强整个地段的通风,收集雨水,并提供多种公共、半公共、私人的活动空间。在户外,不同大小的绿地可以在很多不同的立面中看到。并且绿地和水渠将在水平立面穿行于建筑之间,从而使得这个地段更加宜人。

The cluster defined here contains both spaces typical to the site and unique spaces determined by specific programs. The cluster contains the main metro stop on the site and the regional cultural center. These particular programs inform an urban edge to the canal that is not found elsewhere in the region, along with various edge conditions that allow pedestrians to interact with water.

Buildings are perforated and terraced to increase ventilation across the site, capture rainwater, and provide a variety of outdoor spaces for public, semi-public, and private activity. Outdoor, green spaces of various sizes found at many elevations can support opportunities for agriculture, horticulture, and community recreation. Canals and green paths also perforate buildings at ground level.

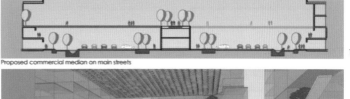

13

14

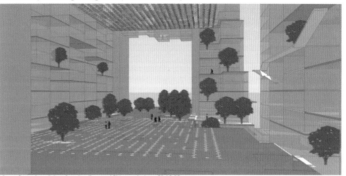

15

16

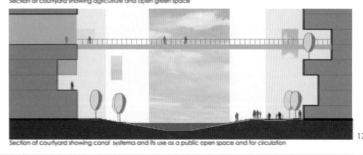

17

12 组团尺度 Group Scale
13 街道剖面 Street Section
14 冬至日庭院透视
  Courtyard Perspective at Winter Solstice
15 夏至日庭院透视
  Courtyard Perspective at Summer Solstice
16 庭院剖面 1 Courtyard Section 1
17 庭院剖面 2 Courtyard Section 2
18 剖面图 Section

18

Longitudinal section-East West 19

生活和办公单元通过优化阳光和自然通风是可变化的。在中国,多代生活在同一住宅是一件非常平常的事情,住宅的面积会随着时间波动。我们所设计的居住和商务单元可以随时间适应居住人数和需求的变化。这样可以让其适应性更强,避免了人们另寻他处,从而继续生活在这一地段。每个单元都是为生活—办公一体化而设计的。

在东西面的商务写字楼,在开窗处使用遮阳板,从而遮挡不必要的阳光以得阴凉。住宅区域设立平面遮阳板为南北向遮阳。我们为所有的单元都设计了开窗口,以用来优化十字向的通风,并为了更充分地获得日光。

Living and working units are transformable over time and optimize sunlight and natural ventilation. In China, where multiple generations living in one household is a common phenomenon, household size can fluctuate over time. Housing and commercial units are designed to be adaptable overtime as occupant numbers and needs change, thus providing spaces that allow individuals to stay and adapt rather than move away. Commercial and office units, with fenestration along the east and west facades, are treated with vertical louvers to shade interior spaces from sunlight. Residential units are treated with horizontal fenestration along their north and south facades. Fenestration is designed in all units to optimize cross-ventilation, and higher volumes are designed to ensure sufficient daylight. Exterior balconies and green rooftops are provided for private outdoor space.

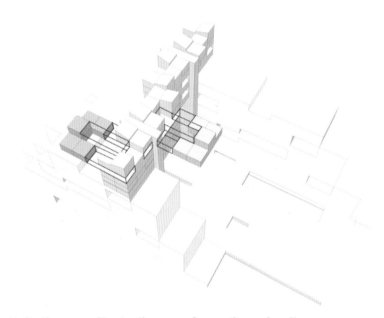

Unit diagram-illustrating configuration of units 20

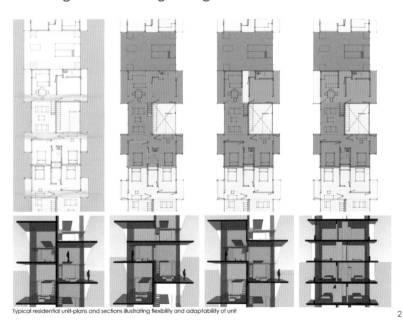

Typical residential unit-plans and sections illustrating flexibility and adaptability of unit 21

19 东西向剖面 East-west Section
20 单元模型 Unit Model
21 典型住宅单元剖面图 Typical Section of Dwelling Unit
22 商业区剖面 Commercial Area Section
23 遮阳百叶/外墙 Sun Louver/External Wall
24 走廊透视图 Hallway Perspective
25 公共空间透视图 Public Space Perspective

Massing

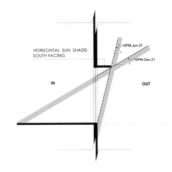
section
水平百叶太阳角度的研究

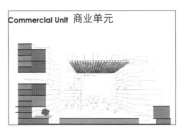
Commercial Unit 商业单元

Perforations

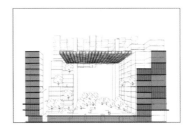
plan
垂直百叶太阳角度的研究

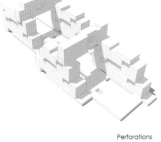

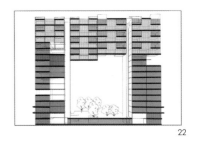

Land uses

Facade study-Horizontal louvres for shading north-south residential facades and vertical louvres for shading east-west commercial/office facades

Green space network

Circulation

View of circulation corridor in commercial/office building showing ideas of vertical open spaces and double facades

View of circulation building showing network of community spaces within the building

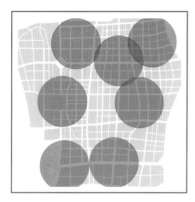

Multiple centers can create greater convenience for user groups and provide a greater variety of neighborhood types.

多中心可以使使用群体感到更加得舒适和多样性。

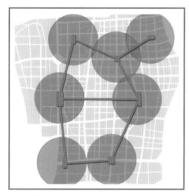

Identified centers are connected with a new metro system, vehicular roads, and the proposed canal system for pedestrians and bicyclists.

被定义的中心将为行人和自行车使用者提供连接地铁的线路、自行车小道以及水渠交通系统的入口。

26

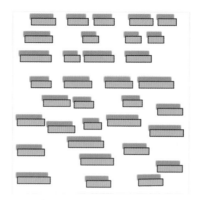

Residential facades face north and south to ensure natural daylight for residents.

住宅南北朝向为了保证日光的渗入。

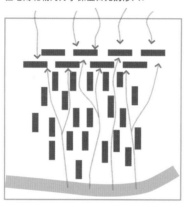

Building masses are oriented north-south at the southern portion of the site to allow south-eastern breezes to penetrate the site in the summer. Buildings oriented east-west along the northern edge block northern winter winds.

地段南部,建筑设计为南北向,如此可使东南风在夏季渗入建筑群。位于地段北边的建筑为东西向用来阻挡冬季的北风。

27

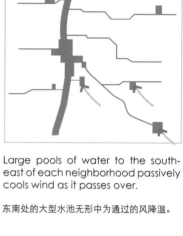

Large pools of water to the south-east of each neighborhood passively cools wind as it passes over.

东南处的大型水池无形中为通过的风降温。

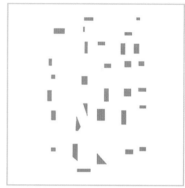

Green spaces on the ground floor are created between buildings and contain public programming.

建筑之间的绿地将作为公共活动场所。

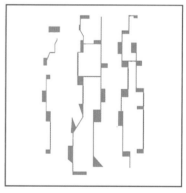

Green paths run perpendicular to the canal system, and intersections create moments for larger public space.

绿色道路将垂直于水渠交通系统,并和公共空间建立联系。

29

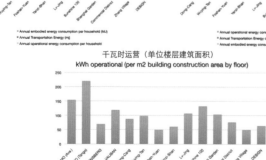
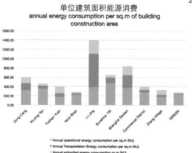

28

26 多中心体系 Multi-central System
27 建筑朝向 Building Orientation
28 能耗分析图 Energy Consumption
29 绿化景观系统 Green Space System

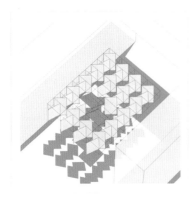

Courtyards can receive a canopy structure of photovoltaic and shade panels. Panels are oriented to let light through during the winter, block light in the summer, and capture solar rays throughout the year.

广场上方将有一个华盖状的遮阳板。遮阳板的设计会使冬天的阳光可以射入，而阻挡夏日的阳光。

Commercial facades are encased with vertical louvers, while residential facades receive a horizontal louver treatment

商务区正面为垂直遮阳板，住宅区则设计为水平遮阳板。

Green spaces at multiple levels and various scales are created to provide for a spectrum of uses, including agriculture, recreation, and horticulture. Vegetation can be used in canals and rooftops to cleanse rainwater.

绿地在不同高度，有不同的光谱作用：农业，娱乐，园艺，而水渠两侧及屋顶的绿化则是为了净化雨水。

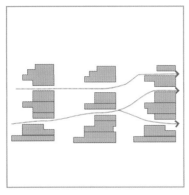

Buildings are perforated at multiple levels to increase natural ventilation through the site.

建筑的缺孔设计以及多阶设计增强整个地段的自然通风系统。

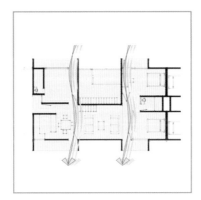

Individual living and working units are designed with floor plans and fenestration to allow for natural ventilation.

设计的独立的住宅和办公单元可由开窗处进行通风。

Vertical connectivity allows for many levels of circulation. Public amenities can thus occur on many levels, decreasing one's reliance on elevators.

垂直连同性可以让阶层之间建立联系，方便残疾人，并减少人们对电梯的依赖。

32

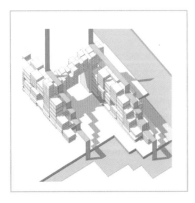

Building terraces capture and filter rainwater, and canals store water to be used at a later time.

梯田状建筑可存储和过滤雨水，储存于水渠之中，以备后用。

30

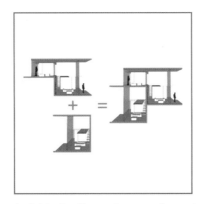

Individual units can be reconfigured over time as household size and membership changes.

独立单元可以随着时间的推延重新组建以适应住宅的面积以及家庭成员变化的需求。

31

30 遮阳百叶 / 外墙 1
Sun Louver/Extenal Wall1
31 遮阳百叶 / 外墙 2
Sun Louver/Extenal Wall2
32 步行系统分析
Pedestrian System Analysis

低碳生活
LOW-CARBON LIFESTYLE

Kui Xue, Caroline Edwards, Jaime Young, Xiaohe Lv, Qi Zhang, Yu Wang, Jessica Lee

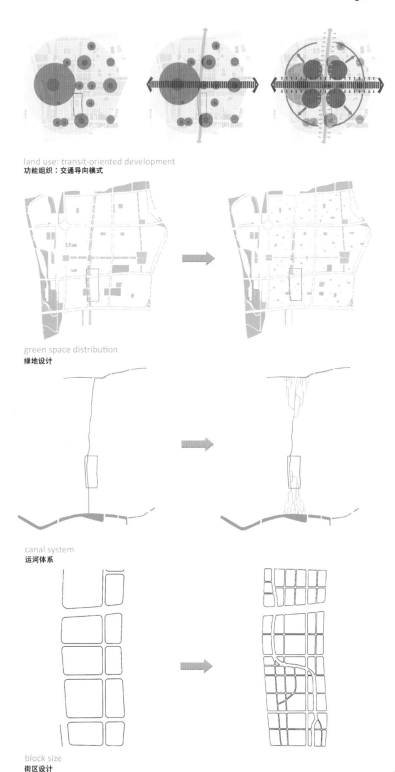

land use: transit-oriented development
功能组织：交通导向模式

green space distribution
绿地设计

canal system
运河体系

block size
街区设计

地段介绍
该地段毗邻城市高铁站及站前商务区之南面，住宅区之北面，南北向运河流经，并有数条地铁线路环绕，交通便捷。

该地段预计将有大量往返CBD的南北向人车流量以及往返轻轨车站的东西向人车流量，两轴线交叉处将形成重要商业人流和公共活动据点。

SITE 4 OVERVIEW
Site 4 resides near the High Speed Rail system and the CBD district with a canal running through the site in south-north orientation.

The site is located in between the highly commercialized CBD on the north end and the residential zones on the south end.

Hence, the site is surrounded by several light rail routes with convenient mobility internally between districts and externally to the historical urban core on the east.

With assumption with large traffic going northward towards the CBD and the light rail station on the east end, an important node will form at the intersection of these two main axis of activities and human flow.

1 设计原则 Design Principles
2 单元分析 Unit Analysis

设计概念说明

城市设计中，我们依托运河水系，将运河水道赋予交通职能，组织各个等级的公共空间，并形成集观览、休憩、娱乐、购物、文教于一体的综合服务序列，同时修复运河水道生境，挖掘其生态保育和涵养功能。在以外部交通网为框架，中央商务区为核心的基础上，组织地段内的配套服务设施，并将不同的功能以单元化的建筑形体镶嵌于整个社区之中，实现生活、生产两位一体的功能混合。在交通系统的设计中，我们采取便捷通达的机动车路网结合丰富的局部立体化非机动车、步行交通体系，使得区域内的交通环境既能够适应高铁站前车流密集的快速行驶需求，又能够营造舒适宜人的步行环境和富有生活感的传统街巷氛围。

围绕节能低碳的理念，我们在社区、组团、建筑单元三个层次采取策略，减少能源耗散，突出环保理念。在住区功能规划上，强调小区域的职住平衡，减少对外通勤交通；在建筑形式的设计上，选取高密度低层数的单体建筑形体，避免高层建筑的运营造成的能源浪费；在生态系统的规划上，重视水系、植物的景观功能和生态功能，积极利用湿地水面、植被板块调节小区微气候，促进自然通风；此外，积极利用太阳能、地源热泵等减少矿物能源的消耗。在设计中，我们通过程序的预测，基本达到了资源集约型社区的目标。

ENERGY/ LIVABILITY CONCEPTS

Our study site of the grid neighborhood in Jinan in many ways informed our new design. Though this existing neighborhood was not designed with energy conservation in mind, it performs very well and also scores high on the livability index. On our first visit, the study neighborhood was immediately appealing as a place to live. The key objectives we developed for our design aim at energy performance but they are no less cognizant of livability as the integration of the two is paramount; for without one, the other becomes irrelevant.

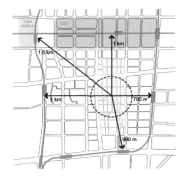

site organization
场地组织

TRADITIONAL GRID 传统网格

3-D GRID 三维网格

	TRADITIONAL	3-D
size (ha)	53	56
residential units	6,879	8,683
pop density (ppl/sq km)	38	47
avg building height	5.5	6.0
FAR	1.7	2.2
energy use/hh (MJ)	74,600	67,600

	TRADITIONAL GRID ENERGY SAVING FEATURES	3-D GRID MODIFICATIONS
BUILDING HEIGHT	majority of buildings ≤ 6 stories ───────→	
LAND USE	ground floor mixed use ───────→	+ vertical mixed use
WALKABILITY	gridded street pattern ───────→ small, permeable blocks ───────→ shaded streets ───────→	+ pedestrian-only streets
GREEN SPACE	vertical ───────→	+ horizontal green space
PUBLIC SPACE	hierarchy of informal spaces ───────→ courtyards ───────→	+ formal spaces

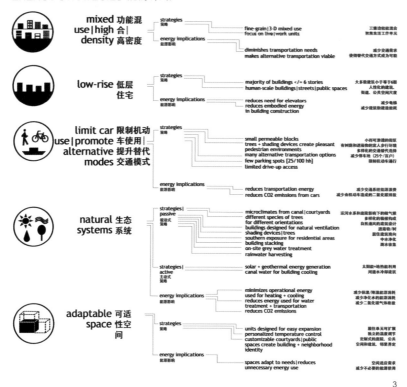

3 能源策略 Energy Strategies

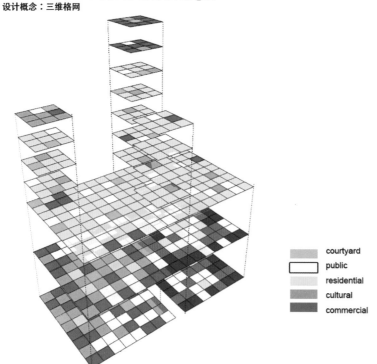

DESIGN CONCEPT: the three-dimensional grid
设计概念：三维格网

- courtyard
- public
- residential
- cultural
- commercial

Mixed use and high density are of utmost importance to our concept. These allow for a walkable environment where residents can fulfill their daily needs within the neighborhood. Low-rise buildings are emphasized because they make for a human-scale atmosphere, require less excavation and embodied energy, and less operational energy in terms of climate control and vertical movement of elevators and utility lines. When designing for swift movement of vehicular traffic, often by default the pedestrian realm suffers which is why we desire to limit access by motor vehicles. Emergency and delivery vehicles are able to circulate on the interior streets but they are otherwise not open to through traffic. Our design disincentivizes private car use but compensates by a rich pedestrian realm that is diverse and convenient and supplemented by a complete multi-modal transit system. The advantages of using natural systems for heating, cooling, air and water are inextricably tied between energy use and livability. Both in architecture and open spaces we designed for optimal operation of passive natural systems. In addition, we are proposing energy generation in several forms in order to offset the increasing energy consumption that corresponds with rising affluence. Our final concept is to make spaces adaptable, both interior private and exterior public spaces. Adaptability for seasonal climate, changes in age and life stage of residents, and transformations in use of spaces over time all were considered. This concept reduces the need for demolition of structures, enhances functionality by allowing for personal choice and customization and provides for a livable environment by allowing for identity of place to emerge.

3 能源策略 Energy Strategies
4 设计概念 Design Concept
5 功能混合 Mixed Use
6 生态系统 Eco-system
7 交通系统 Transportation

区域／社区尺度

首先我们对现有的土地利用规划进行了一些调整，使得土地利用能够允许更加细化的混合使用。其次我们取消了一些大型的公园绿地和高速路边的绿化廊道，将绿地细分为较小的植被板块，进一步区分绿化空间的等级，依次设置中央公园、社区公园、街心花园以及建筑单元尺度的绿化庭院。再次我们尝试改变运河水道布局，在流进中央商务区时保持原有形态，但在其流入居住区时，分流为更加细小的河道，以期界定出符合我们设计理念的格网式街区的空间，并使其尺度与周围环境相统一。

地段西北部毗邻中央商务区，具有重要的居住职能。我们改变了点状聚集的商业空间模式，试图将商业空间分散至整个地块，使各个居住组团能以便捷的步行方式抵达。日常购物和生活服务设施更是尽量分散设置在邻里中，同时以公交线路、水上出租车线路、道路网络将这些居民点与中央商务区、文教设施、地铁、高铁车站相连接，提供更多的选择。

组团尺度

主要的节能设计在组团尺度上显现出来。建筑配合夏季西南风，以天然风在不同尺度大小的内院创造微气候来提供天然制冷。位于二层三层的连接天桥提供居民更快速便捷及安全的行动环境，也同时活络一层以上的商业文化活动。在树木选择上，会在建筑南面种植落叶乔木，夏天遮蔽太阳，冬天让冬阳照射建筑提供天然制暖。建筑北面选择种植常青树木，夏天为行人提供树荫，冬天则能阻挡西北风。地面设计则会选择通透性建材，提供雨水采收再利用功能，同时减低热岛效应。水道的多尺度设计不仅能提供天然的净水功能，提供优良的微气候，也同时提供节能的公共交通运输。

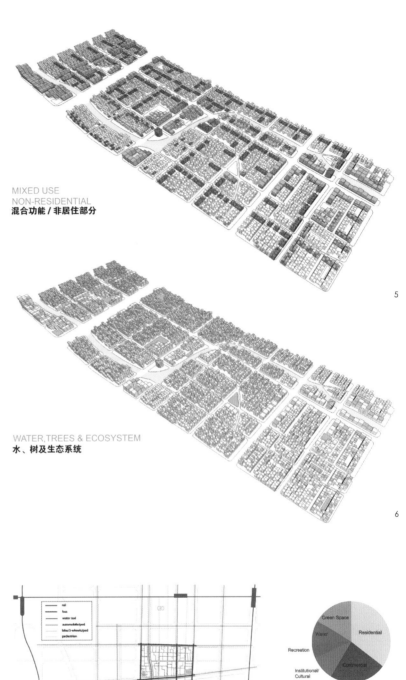

MIXED USE
NON-RESIDENTIAL
混合功能／非居住部分

WATER, TREES & ECOSYSTEM
水、树及生态系统

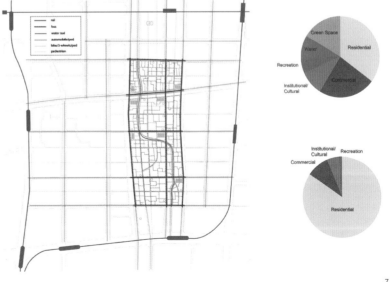

solar panels
sunlight collecting
太阳能板

roof gardens
rain harvest
空中花园

trees & ecosystem
deciduous and evergreen trees are planted on different locations
常绿树及落叶树

circulation system
ground pedestrian circulation
medium level pedestrian loop
car circulation / parking
connected to water transit
步行系统
车行系统
水运

place making
creating meaningful places for different groups of people
场所营造

non-residential
mixed-use
混合功能 非居住部分

cluster overall aerial view
组团总体轴测图

AREA/NEIGHBORHOOD SCALE

The existing land use master plan we found to be suitable with a few alterations. The land uses shown should allow for a finer-grain mix of uses such that a parcel may contain multiple land uses and a single building will do the same vertically. Along with this we propose that the green spaces be broken into smaller chunks and disseminated. Rather than a few large parks and green swaths along the highway, we envision a hierarchy with some large parks, neighborhood parks, pocket parks, and green courtyards at the building and even unit level. Our third proposed change is with the layout of the canal. It will remain intact in the central business district but then distribute into smaller canals as it flows through the residential areas. In this way it acts as an organizing principle among the blocks of the grid, defining spaces as it moves through. Its scale and intensity is thus allowed to adapt to or inform its surrounding context.

Site four in the context of the area plays the role of a residential neighborhood amidst intense commercial to the north and lighter commercial to the south. We felt that commercial activity should be integrated throughout rather than concentrated at nodes that are beyond walking distance from residential areas. The everyday goods and services are all provided within the neighborhood at smaller nodes while the central business district, major institutions, rail stations, etc. are linked to our site by bus, water taxi, and a network of roads that create multiple options for mode choice. The blocks of the neighborhood are small and permeable such that walking is made convenient, thereby reducing the need for transportation in general. Low-rise buildings lead to energy savings at both the embedded and operational levels and are designed along with the spaces between to allow for airflow, light, shading, and wind protection.

8 区域 / 社区、组团尺度分析
Region/Community/Group Scale Analysis

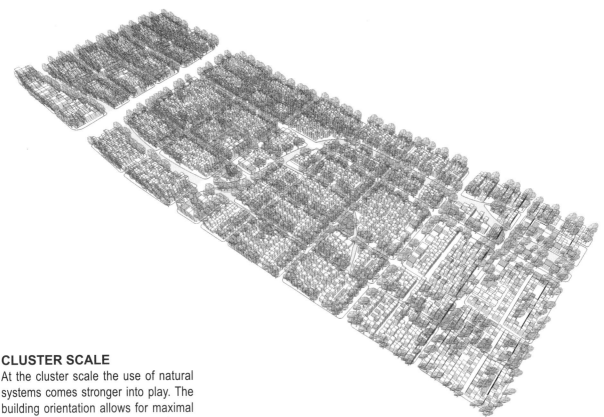

CLUSTER SCALE

At the cluster scale the use of natural systems comes stronger into play. The building orientation allows for maximal ventilation while the courtyards on various levels are designed to provide a cooled microclimate to the interior rooms. Circulation at the ground, 2nd and 3rd levels means more options for efficient pedestrian flow so when moving between buildings, one only must go to the floor with the path rather than to the ground each time. The tree and vegetation scheme consists of evergreen trees on the north sides to block winter winds and deciduous trees on the south to block intense summer sun while allowing for winter sun when they lose their leaves. Groundcovers and shrubs diminish the amount of heat-absorptive surfaces and further contribute to cooler microclimates and while also reducing impervious surfaces for the sake of runoff. Pavement and building roofs are shaded by tree canopy to reduce heat-island effect. The canal permits greater airflow by creating a channel between buildings and provides cooled air to adjacent buildings. In the residential areas where public space is less commercial, adjacent buildings are served by greywater filtration zones that abut the canal and add defining character.

ENERGY STRATEGIES @ NEIGHBORHOOD SCALE

Mixed-use/high-density
o   Fine-grain, 3-D mixed use

Low-rise
o   Human-scaled neighborhood
o   Integrated building/street life

Limit car use/promote alternative modes
o   Limited parking
o   Motor vehicles relegated to peripheral streets
o   Network of shared vehicles
o   Buses on all major streets
o   Water taxi to link CBD and southern transit stop
o   Internal streets for efficient bicycle circulation
o   Permeable blocks for further pedestrian access
o   Gridded street pattern creates connectivity

Natural systems

o   Building massing allows for light, airflow, wind protection
o   Tree canopy designed to shade streets and building surfaces, reduce heat island effect
o   Microclimate around canal edge
o   Network of geothermal wells

Adaptable space

o   Open space programming dictated by neighborhood

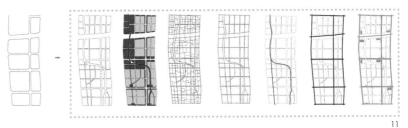

9 俯视图 Top View
10 社区节能策略 Energy Strategies at Neighborhood Scale
11 交通分析 Traffic Analysis

055

12 组团平面图 Building Group Plan
13 组团俯视图 Building Group Top View
14 组团设计 Building Group Design
15 剖面图 Section

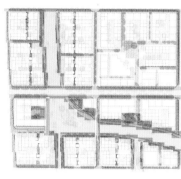
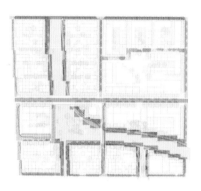

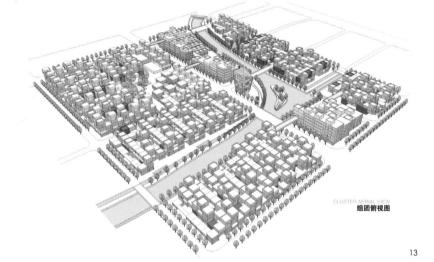

组团平面图 CLUSTER PLAN 12

组团俯视图 CLUSTER AERIAL VIEW 13

14

15

东西剖面 SECTION THROUGH WEST TO EAST

南北剖面 SECTION THROUGH SOUTH TO NORTH

16 河岸行道透视图 Riverbank Walkway Perspective

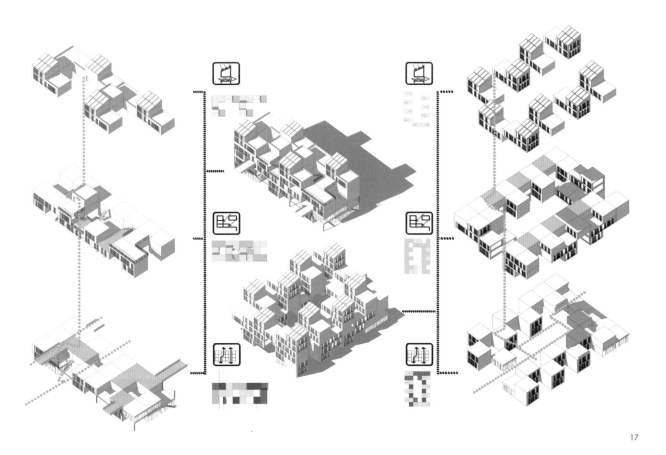

16 河岸行道透视图 Riverbank Walkway Perspective
17 建筑单元组合模式 Building Unit Composite Mode

057

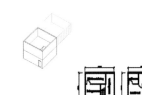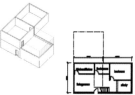

UNITS LAYOUT
建筑单元设计

FENESTRATION TYPES
建筑外立面类型

单元尺度的设计理念

居住单元的设计具有可变性，可适应不同生活方式及不同规模家庭的需求。其中，核心单元为 SOHO 住宅，满足一系列家庭办公的需求。核心单元通过不同组合可以满足大家庭的需要。所有的单元模数的设计基于柔性设计理念，易于扩建和相互联系。同时，基本的居住单元结合整合的功能、社区空间、绿化空间和交通系统以立体化的联系与沟通组织在建筑群中。结合垂直绿化体系，建筑物的底层庭院绿化体系为居民创造了舒适的冬暖夏热的微气候。

UNIT SCALE

The units are designed to be easily adaptable to meet the needs of different lifestyles and changing family sizes. The core unit is a small-office/home-office ("SOHO") live/work unit that can accommodate a range of home businesses. Modifications of the core unit form units for larger families. All unit sizes are designed to be flexible, expanding and contracting as necessary. Units are arranged into buildings that integrate uses, community space, green spaces, and circulation three dimensionally. In addition to vertical green spaces, the buildings feature ground-level courtyards, which form cooling microclimates and are adaptable to the needs and desires of the building inhabitants.

ENERGY STRATEGIES @ UNIT SCALE

 Mixed-use/high-density:
  Groundfloor and vertical mixed use

 Low-rise:
  Majority of buildings ≤ 6 stories

Natural systems:
  Southern orientation, trombe walls for passive heating
  Orientation to take advantage of cooling winds in the summer
  Building openings to allow for wind flow
  Building stacking
  Courtyards and vertical green spaces create microclimates
  Fenestration options maximize passive heating & cooling
  Rainwater collection & onsite greywater filtration
  Geothermal heating
  Integrated solar pv and hot water systems
  Canal water used for building cooling

 Adaptable space:
  Units easily expandable
  Personalized temperature control
  Fenestration options
  Customizable courtyards/public spaces

能源策略

功能混合 / 高密度
  南向的主体墙利用太阳能被动取暖
  建筑物朝向利用夏季的冷风降温
  建筑空间利于空气流循环
  建筑分层
  庭院和垂直绿化空间营造微气候
  多样化的开窗选择增加被动式加热和降温机会
  雨水收集和就地中水净化
  地源热泵
  集中式的太阳能热水系统
  运河水用作建筑物降温
可适性空间
  易于扩建的单元
  独立温控系统
  开窗选择
  定制化的庭院和开放空间

18 建筑单元设计 Building Unit Design
19 建筑外立面类型 Building Facade Types
20 能源策略 Energy Strategies
21 院落透视图 Courtyard Perspective
22 能源分析 Energy Consumption Analysis

VIEW OF TYPICAL LINEAR COURTYARD
院落透视图
21

能源总论

该设计中，每个家庭一年预计能源使用为 67600 MJ。利用混合功能高密度底层，严限汽车使用，提倡其他节能交通工具，利用天然制冷制热系统以及提供活用空间等设计元素，该设计是个节能宜居的现代规划。值得一提的是，该设计大概只耗用近期在济南的高层开发案一半的能源。再者，如与国际上的优良设计典范相比，该设计的使用操作上的消耗能量（以平方米为单位）是低于位于瑞典的 Bo01 方案以及位于德国的 Vauban 方案。在能源分析中，使用操作上的能源使用是该设计的主要能源使用，占据每个家庭能源使用的百分之八十。建筑融入能占百分之十三，而交通运输能占百分之七。以使用操作能来说，其中将近一半都用于制热上，如暖气等电器用品都是不包括在设计范围内的，而百分之四十二则是其他能源使用综合。总结来说，制冷或是电灯采光等能源使用与制热用电相较之下属次要的耗能因素。

ENERGY ANALYSIS 能源分析

PROFORMA RESULTS 形势结果

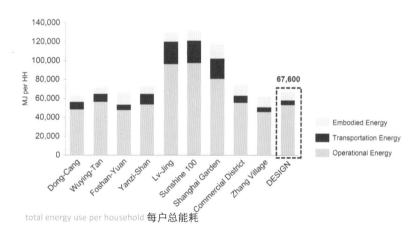

total energy use per household 每户总能耗

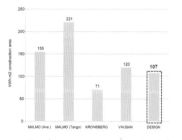

operational energy use per m² constructed area
每平方米建筑面积运营能耗

household operational energy use
家庭运营能源使用

22

23 策略分析
Energy Summary
24 透视图
Perspectives

 1 mixed use/high density

 2 low rise

3 alternative transportation options
4 small, permeable blocks
5 few parking spots, limited car access
6 shaded streets, pleasant pedestrian environments
7 microclimates
8 different tree species for different orientations
9 buildings designed for passive heating/cooling
10 onsite rainwater harvesting/greywater treatment
11 geothermal
12 combined solar hot water/pv systems
13 canal water for building cooling
14 units designed for expansion
15 customizable courtyards/public spaces

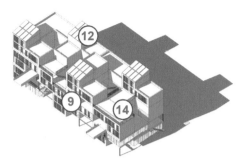

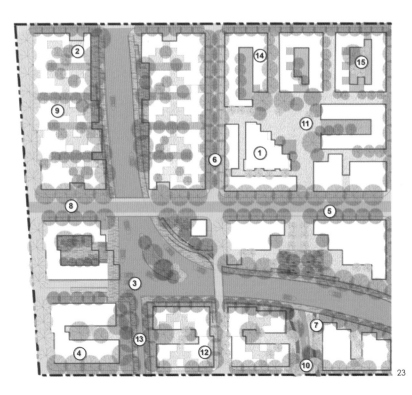

ENERGY SUMMARY

The estimated annual energy use per household for the 3D Grid design is 67,600 MJ. Using the energy strategies of low-rise buildings, mixed use/high density, limited car use and encouragement of alternative modes of transportation, integration of natural systems, and creation of adaptable space, the 3D Grid is a livable, modern development that uses less energy per household than six of the nine neighborhoods studied in Jinan. Most notably, the 3D Grid uses roughly half the energy used by the recent, high-rise developments in Jinan. The 3D Grid also performs favorably compared to international best practices for energy efficient urban design, with an operational energy per square meter constructed area that is lower than Bo01 in Malmo, Sweden and Vauban in Freiburg, Germany. Energy use for the 3D Grid is dominated by operational energy uses, which account for 80% of total household energy consumption. Embodied energy and transportation energy account for 13% and 7%, respectively, of total household energy use. Nearly half of household operational energy use is devoted to heating, with another 42% used for "other uses", such as electric appliances, which are not impacted by design.

2011

济南
西部新城 | JINAN
NEW TOWN
清洁能源城市 | CLEAN ENERGY CITIES

清洁能源城市
CLEAN ENERGY CITY

1 地段区位 Location in Jinan New Town
2 总平面图 Site Plan
3 组团平面图 Cluster Ground Floor Plan

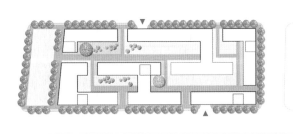

065

年家庭能耗
Annual Operational Household
Energy Consumption by Use

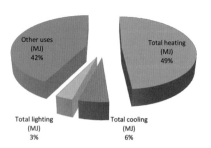

公共区域户均能耗
Per Household Common Area
Energy Consumption

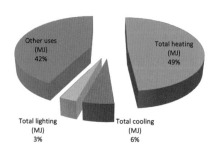

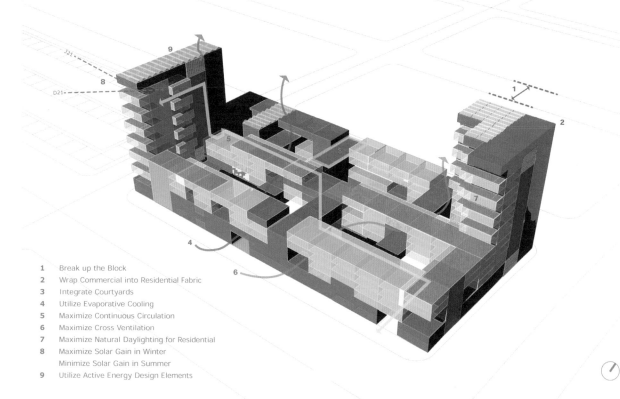

1 Break up the Block
2 Wrap Commercial into Residential Fabric
3 Integrate Courtyards
4 Utilize Evaporative Cooling
5 Maximize Continuous Circulation
6 Maximize Cross Ventilation
7 Maximize Natural Daylighting for Residential
8 Maximize Solar Gain in Winter
　Minimize Solar Gain in Summer
9 Utilize Active Energy Design Elements

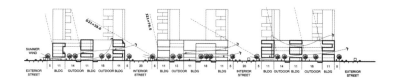

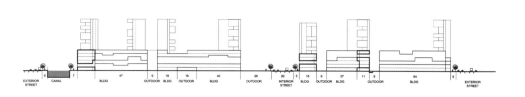

4 能源 + 宜居性策略
Energy+Livability Strategies
5 西向剖面图
Section Looking West
6 北向剖面图
Section Looking North
7 商业连续
Continuous Commercial Program
8 居住连续
Continuous Residential Program
9 组合连续
Continuous Composite Program
10 塔楼不同类型平面
Different Plans of Tower

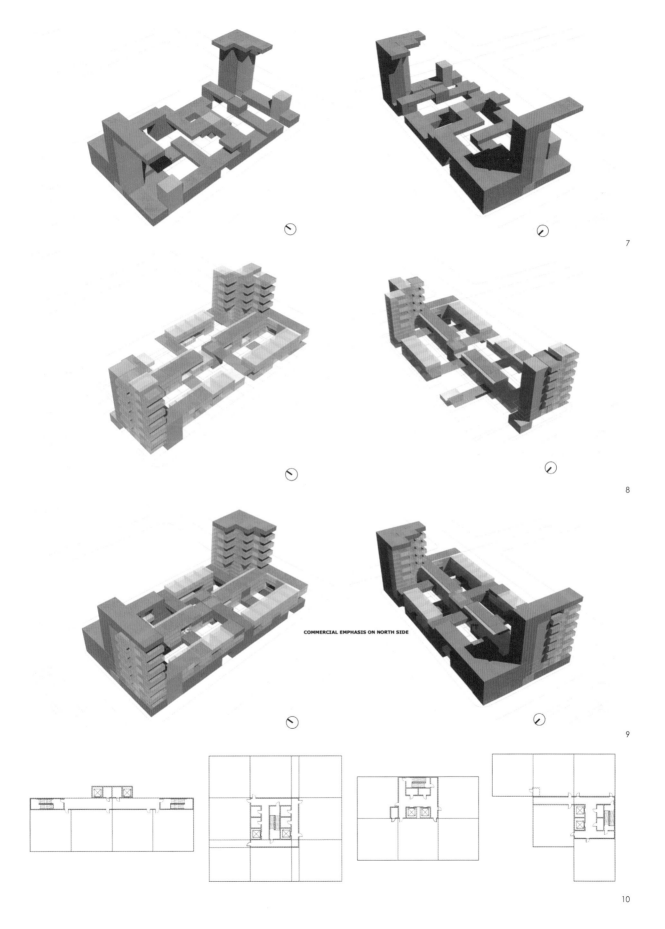

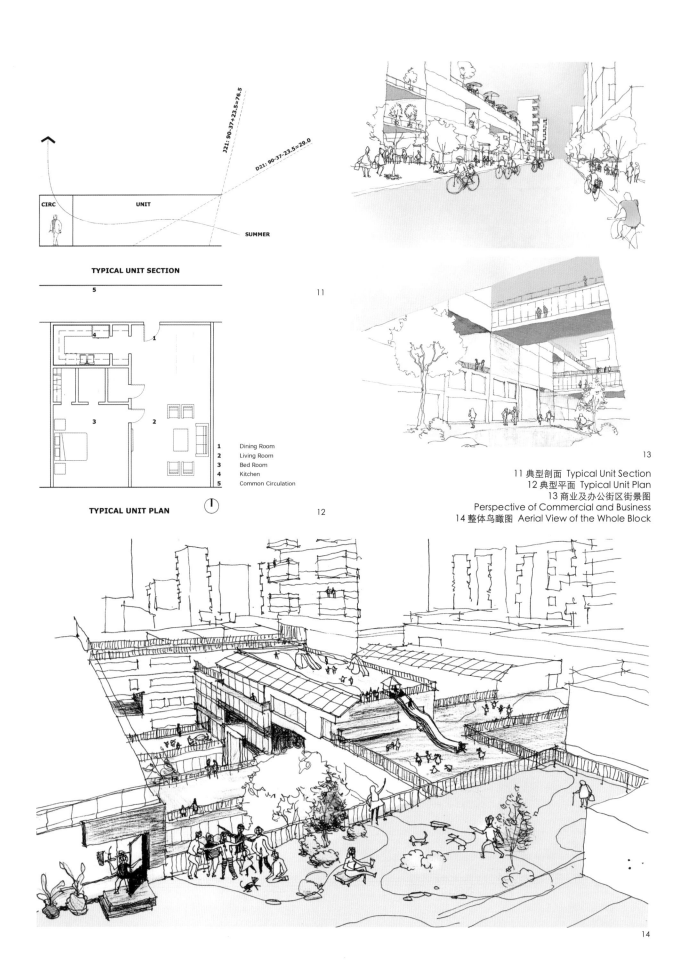

TYPICAL UNIT SECTION

$J21: 90-37+23.5=76.5$
$D21: 90-37-23.5=29.0$

CIRC | UNIT | SUMMER

TYPICAL UNIT PLAN

1 Dining Room
2 Living Room
3 Bed Room
4 Kitchen
5 Common Circulation

11 典型剖面 Typical Unit Section
12 典型平面 Typical Unit Plan
13 商业及办公街区街景图 Perspective of Commercial and Business
14 整体鸟瞰图 Aerial View of the Whole Block

空中城市
STALAGMITE CITY

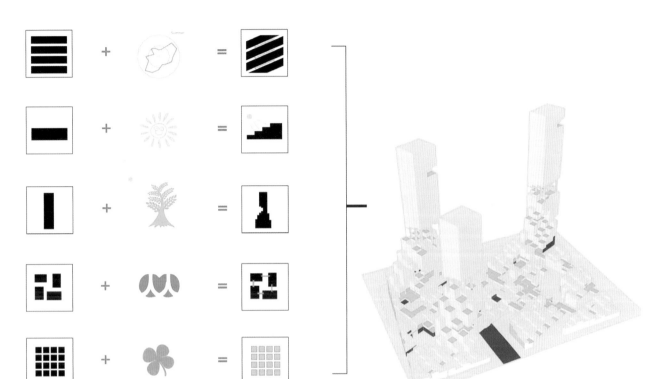

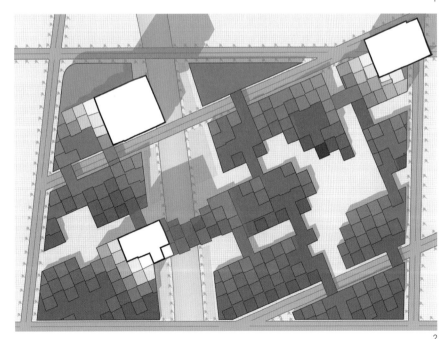

1 要素叠加 Integration of Functions and Spatial Forms
2 总平面图 Master Plan

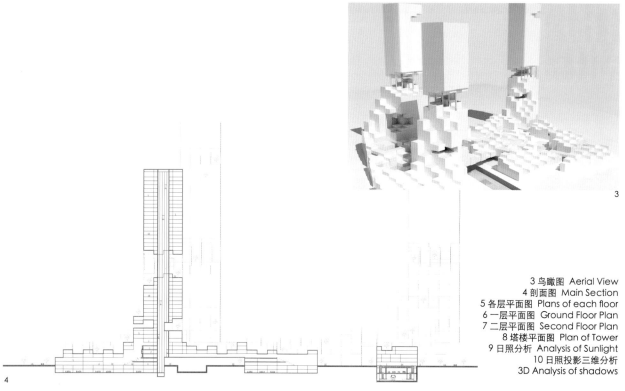

3 鸟瞰图 Aerial View
4 剖面图 Main Section
5 各层平面图 Plans of each floor
6 一层平面图 Ground Floor Plan
7 二层平面图 Second Floor Plan
8 塔楼平面图 Plan of Tower
9 日照分析 Analysis of Sunlight
10 日照投影三维分析 3D Analysis of shadows

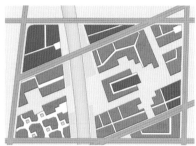
一层平面 FLOOR 1

二层平面 FLOOR 2

三层平面 FLOOR 3

四层平面 FLOOR 4

五层平面 FLOOR 5

六层平面 FLOOR 6

七层平面 FLOOR 7

八层平面 FLOOR 8

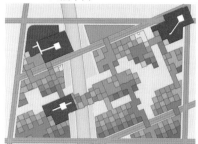
九层平面 FLOOR 9

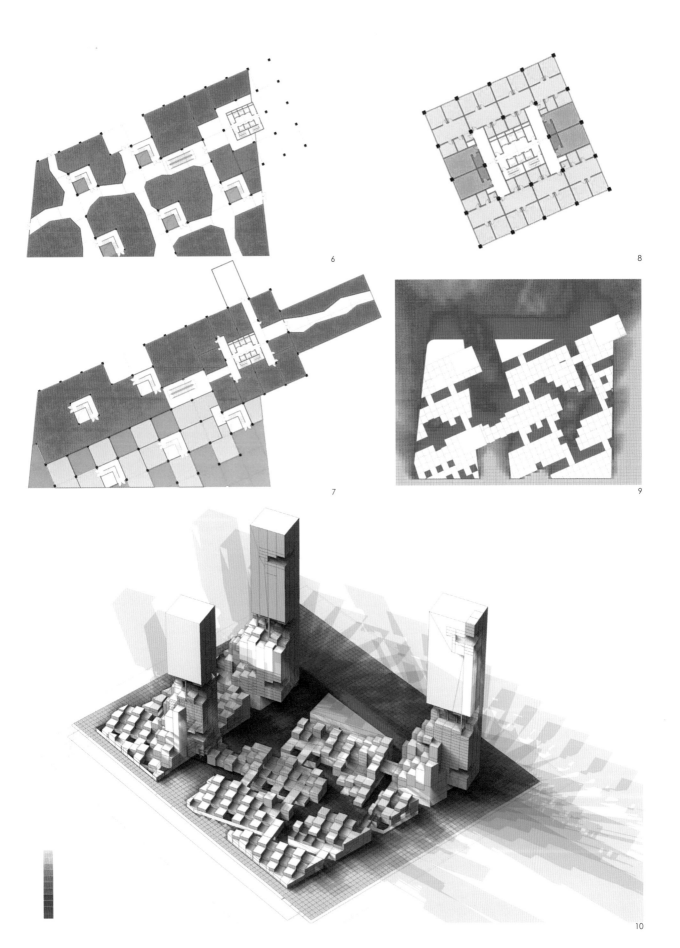

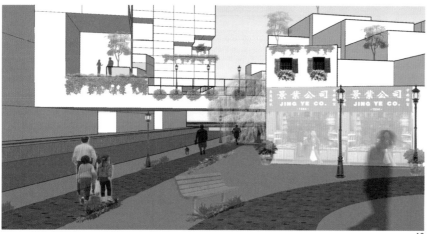

11 多层次庭院
Multi-level Courtyards
12 屋顶花园鸟瞰图
Aerial View of Roof Gardens
13 底层商业透视图
Perspective of Commercial at Bott
14 立面图
Main Facade

清洁能源城市
CLEAN ENERGY CITY

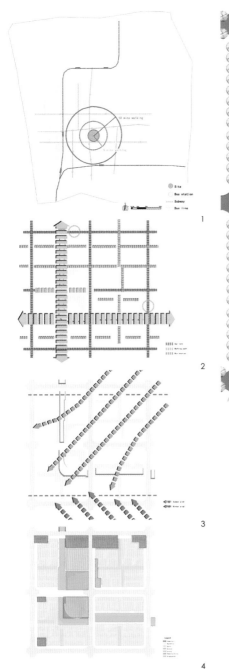

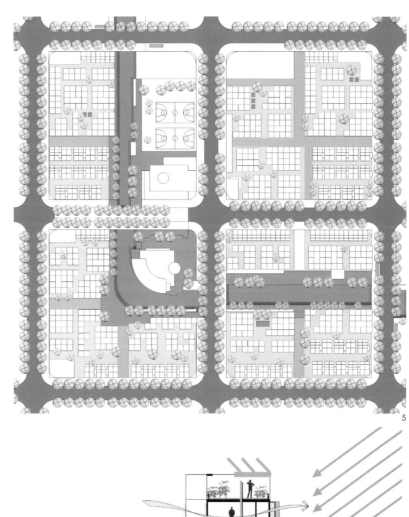

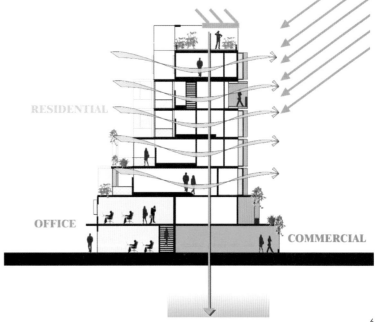

1 区位图 Location Analysis
2 道路交通分析 Classification of Roads
3 季风分析 Anlysis of Monsoons
4 功能分区 Function Zoning
5 总平面图 Master Plan
6 日照通风分析 Analysis of Sunlight and Ventilation

073

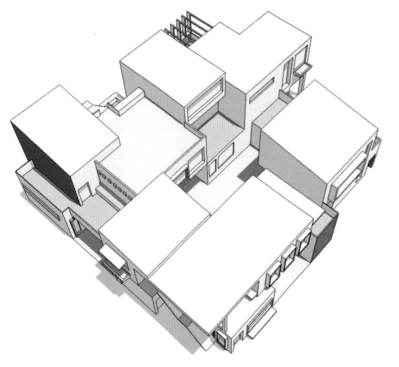

能源高效利用策略：城市尺度

1. 高密度、扁平化
A. 所有单元都可以通过楼梯和廊道到达
B. 人体尺度设计保证步行可达
2. 邻里间步行可达
A. 城市快速交通站点附近邻里空间主导
B. 街区间的步行道网络
C. 沿河的散布道和自行车道
3. 功能混合利用
A. 所有商业零售均混合到居民楼中
B. 公共空间（包括开放和封闭的）通过邻里组团综合在一起
C. 学校和其他公共设施散布在居住空间中

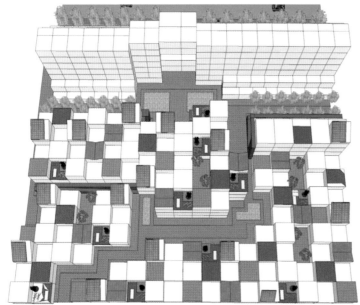

Energy Efficient Design Strategies: City Scale

1. High-Density, Low-Rise
A. all units are accessible by stairs and ramps
B. human-scale design enhances walkability
2. Walkable Neighborhoods
A. neighborhood oriented around bus rapid transit (BRT) stations
B. network of pedestrian paths through block
C. pedestrian promenade and bikeway along river
3. Mix of Uses
A. all commercial retail integrated into residential buildings
B. public space (both open and enclosed) incorporated throughout neighborhood clusters
C. schools and other public facilities interspersed with residential uses

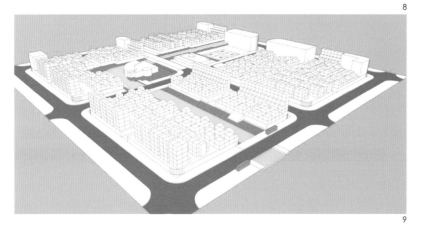

7 组团单元
Cluster Unit
8 模数化屋顶平面
Modulization of Roof Plan
9 整体高度控制
Integral Control of Height
10 屋顶太阳能发电利用
Solar PV into Roof Design
11 不同类型组团
Different Kinds of Clusters
12 日照分析
Analysis of Sunlight and Shadows

能源高效利用策略：自然元素
4. 太阳能
A. 建筑的朝向获得南部采光最大化
B. 建筑上的遮阳元素使居民在夏季保持凉爽
C. 棋盘模数设计使得建筑的阴影和自然采光可以交错实现
5. 风能
A. 场地上对角线的风向流动使得夏季通过空气流通带来微风
B. 较高的建筑庇护步行小径和开放空间免受冬天寒风
C. 单元在南侧和北侧立面均开窗，使得空气流通

Energy Efficient Design Strategies: Natural Elements

4. Sun
A. building orientation maximizes southern exposure
B. shading elements on buildings keep residences cool in summer
C. checkerboard design allows for mutual building shading and natural light
D. taller buildings shade streets and open spaces
E. solar PV incorporated into roof design
5. Wind
A. diagonal wind corridor across site allows for summer breezes
B. taller buildings shild paths and open spaces from winter winds
C. units have windows on south and north facades allowing for cross-ventilation

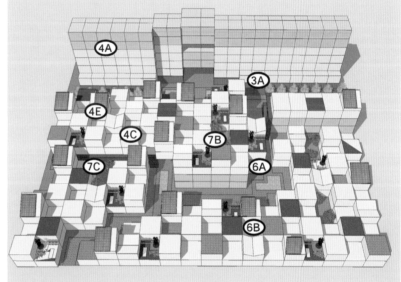

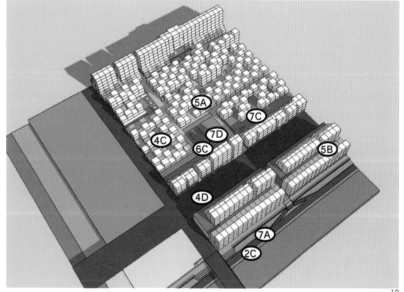

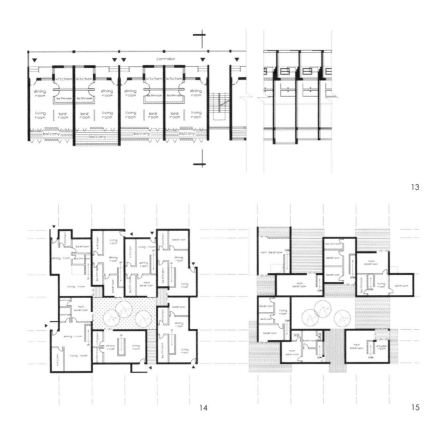

13 高层户型平面与剖面图
Unit Plan and Section for Tall Building
14 首层平面图
Ground Floor Plan
15 二层平面图
Second Floor Plan
16 北侧立面
North Facade
17 冬季季风流动
Winter Monsoon Flow
18 夏季季风流动
Summer Monsoon Flow

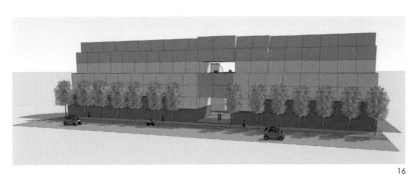

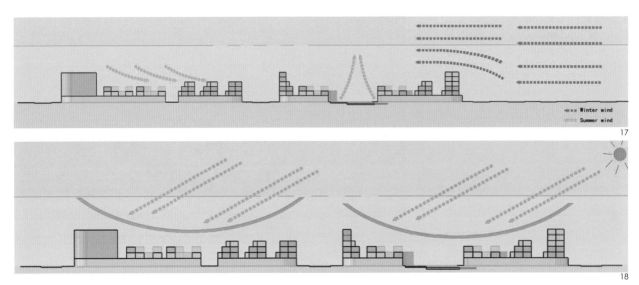

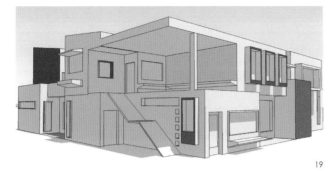
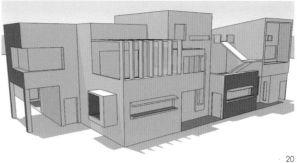

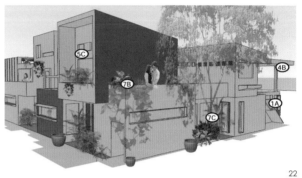

19 组团单元 A Cluster A
20 组团单元 B Cluster B
21 高层建筑南侧立面
South Facade of Tall Building
22 组团单元绿植分布
Distribution of Plants in One Unit
23 沿河自行车道及步行道透视
Perspective of Pavement and Bicycle Path
24 社区公共空间透视
Perspective of Public Space of Neighbourhood
25 步行商业街透视
Perspective of Pedestrian Commercial Street

2012

济南
城中村
清洁能源城市

JINAN
URBAN VILLAGE
CLEAN ENERGY CITIES

农业—文化—生活实验室
AGRI/CULTURE LIVING LAB

Alexis Wheeler, Cecilia Ho, Luan Bo, Matthew Bunza, Xie Yang

CHINA: CULTURE OF FOOD 中华饮食文化

SHANDONG: REGIONAL CONNECTIONS 北京和上海间的中转站 **JINAN**: CELEBRATION OF WATER

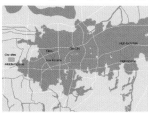

DIKOU: CITY OVERTAKING FARMS 被城市压倒的农业 **JINAN**: SUPPLY / CONSUMPTION

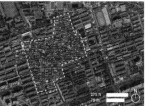

1 设计场地
Aerial with Systems Overlay
2 背景分析
Background

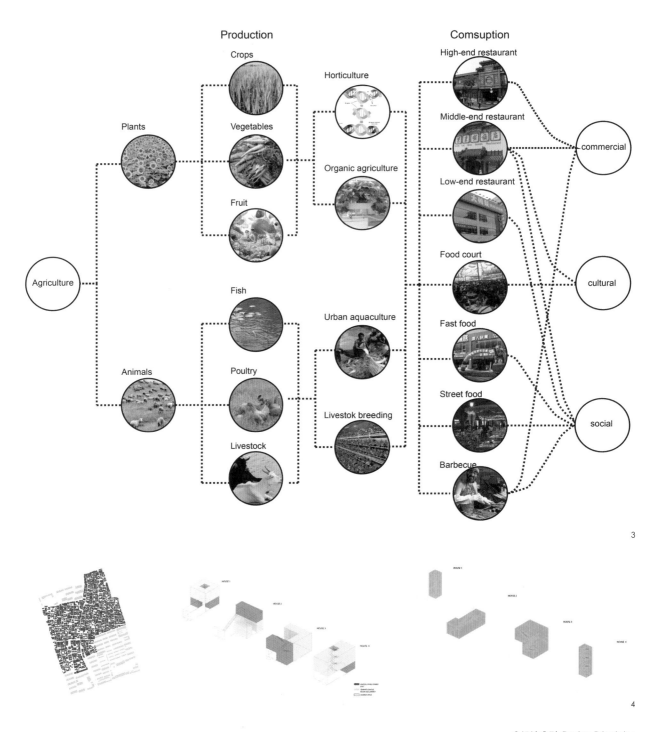

3 设计准则 Design Principles
4 堤口研究 Dikou Research
5 总平面图 Master Plan

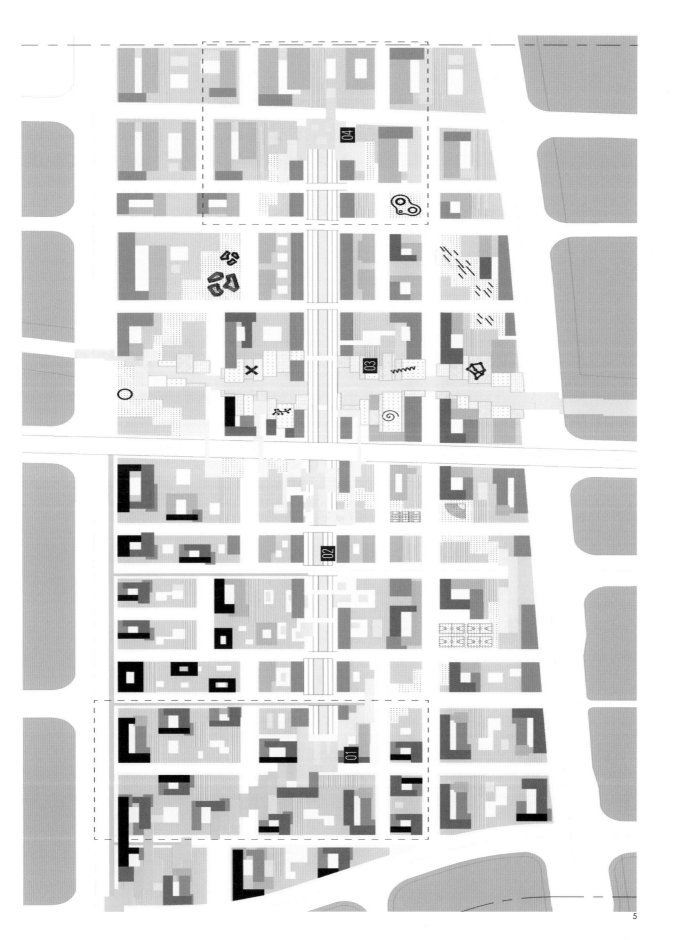

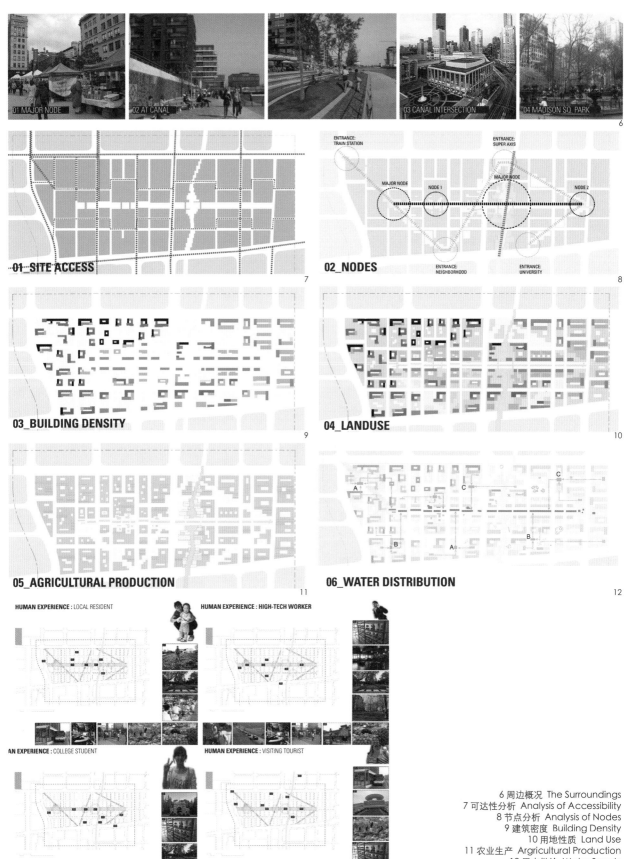

6 周边概况 The Surroundings
7 可达性分析 Analysis of Accessibility
8 节点分析 Analysis of Nodes
9 建筑密度 Building Density
10 用地性质 Land Use
11 农业生产 Argricultural Production
12 用水供给 Water Supply
13 使用者感受 Human Experience
14 组团分析 Analysis of Clusters

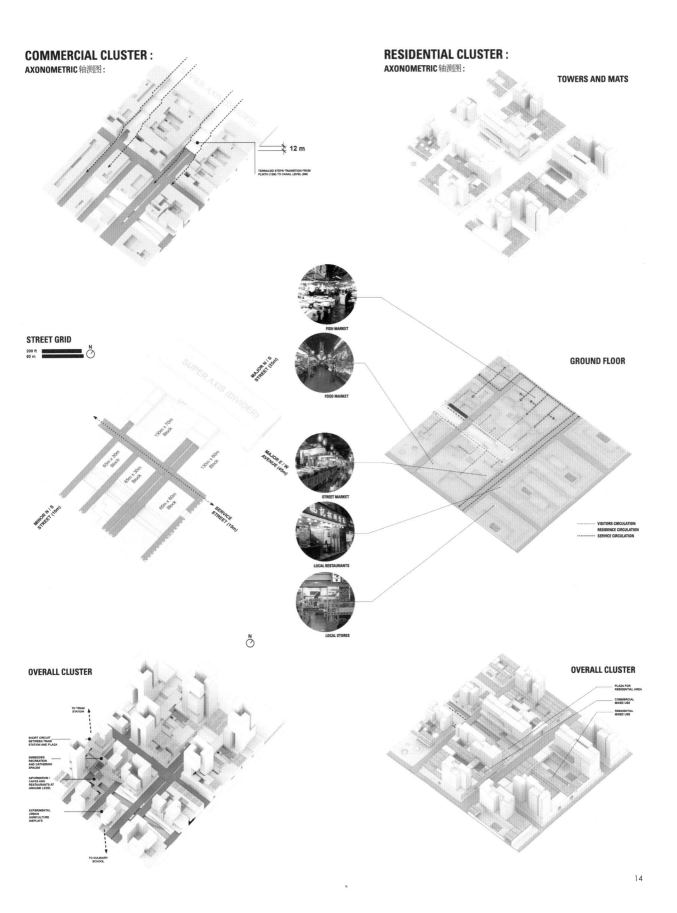

15 商业组团透视示意 Commercial Cluster
16 居住组团透视示意 Residential Cluster
17 塔楼裙房单元分析 Units of Tower Podiums
18 组团剖面分析示意 Section of Cluster

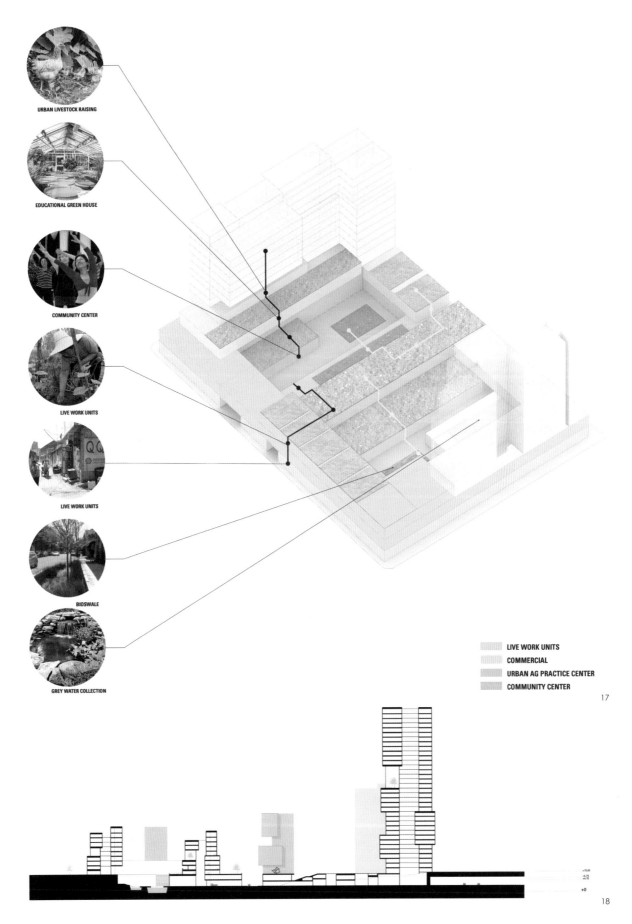

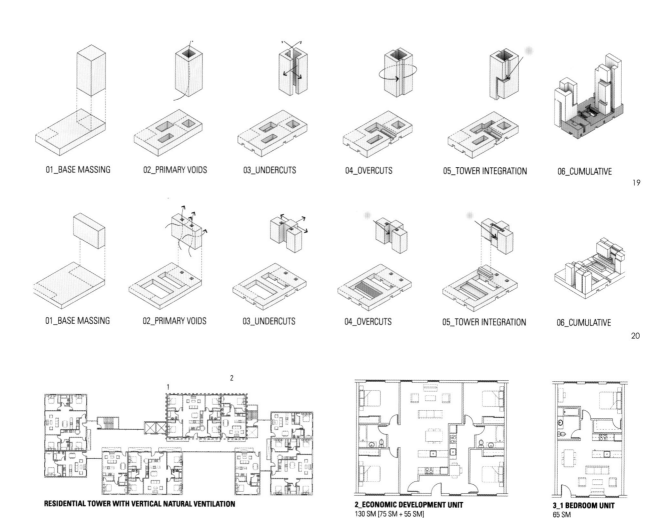

| 01_BASE MASSING | 02_PRIMARY VOIDS | 03_UNDERCUTS | 04_OVERCUTS | 05_TOWER INTEGRATION | 06_CUMULATIVE |

19

| 01_BASE MASSING | 02_PRIMARY VOIDS | 03_UNDERCUTS | 04_OVERCUTS | 05_TOWER INTEGRATION | 06_CUMULATIVE |

20

RESIDENTIAL TOWER WITH VERTICAL NATURAL VENTILATION

2_ECONOMIC DEVELOPMENT UNIT
130 SM [75 SM + 55 SM]

3_1 BEDROOM UNIT
65 SM

21

LANDLORD IN MAIN UNIT LEASING SRO'S WITH SHARED FACILITIES

LANDLORD IN MAIN WITH EXTENDED FAMILY IN ADJACENT UNIT

EXTENDED FAMILY MERGING UNITS INTO SINGLE UNIT

22

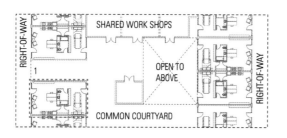

1_LIVE/WORK UNIT
70 SM

23

19 商业塔楼概念
Concept of Commercial Tower
20 居住塔楼概念
Concept of Residential Tower
21 居住塔楼单元示意
Units of Residential Tower
22 单元经济适应性提升
Economic Development Unit Flexibility
23 裙房居住 / 工作单元示意
Live/Work Units in Mat Slab
24 自然通风策略
Natural Ventilation Strategies
25 综合水系统
Integrated Water System

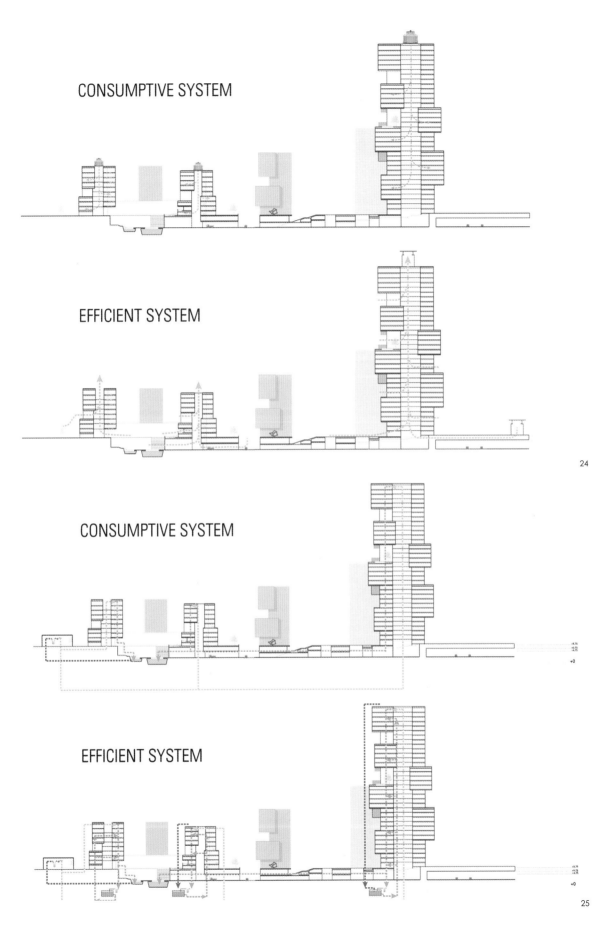

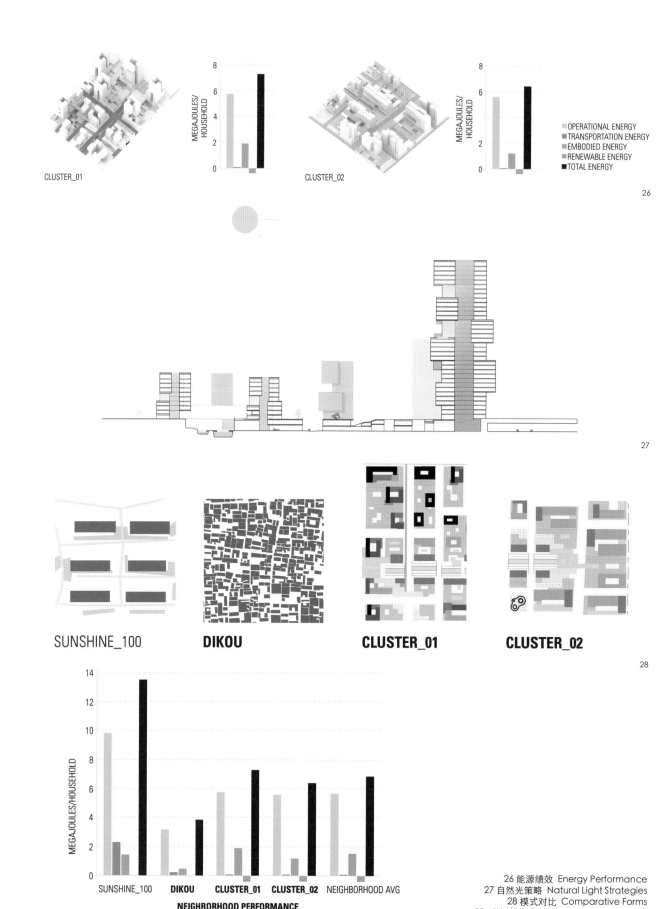

26 能源绩效 Energy Performance
27 自然光策略 Natural Light Strategies
28 模式对比 Comparative Forms
29 对比绩效 Comparative Performance

知识庭院：人与自然、人与人之间和谐交流的场所
COURTYARDS OF KNOWLEDGE: CONNECTING PEOPLE, PLACES AND IDEAS

Andrew Turco, Feifei Feng, Francisco Humeres, Minjee Kim, Xin Xiu Chang

济南西区：合作互动的中心枢纽

由于处在京沪高铁的大约中点的位置，济南承担着上海、北京、青岛之间的交通中转站的作用。这座中国历史上重要的商业中心城市现在被定位成周边城市的会议交流中心——在这里，人群和商贸得以汇聚合作、相互影响，并在越来越依赖智慧创新的经济平台上交流思想。我们将这一区域看作是该城市的科教创新区——也和济南作为中国东海岸城市汇聚地的新角色有所契合。

WEST JINAN: CENTER OF COLLABORATION AND EXCHANGE

As a result of its relatively central position along the new Shanghai-to-Beijing High Speed Rail Line, Jinan is poised to become the central hub between Shanghai, Beijing, and Tsing Tao. This city that has historically been one of China's major commercial centers is now positioned to become a meeting point for those coming from these surrounding cities – the place where people and business converge to collaborate, interact, and share ideas in an economy increasingly dependent on knowledge and innovation.

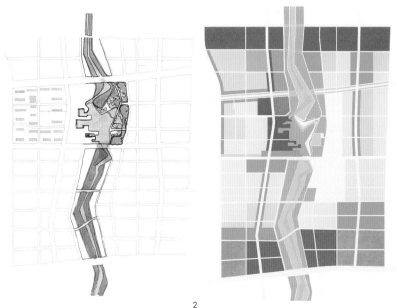

1 鸟瞰图 Aerial View
2 贯穿南北邻里的河流脊柱
A Neighbourhood Spine from North to South
3 用地规划 Land Use Planning

water
rough green
park
residential - low density
residential - moderate density
residential - high density
mixed-use
university
office and R&D
commercial

091

老济南的启示：再访旧商业区的有机路网

济南北部的商务区的方格路网体系由德国人建于1897年，这些主干道传递着区域之间的交通，但是和中国其他新兴发展模式相对比，这些主干道并没有造成使社区分割的不能通过的障碍。除了这些主要通道，还有各种各样的次级道路和在街区内部的小径。这些捷径减少了家庭到附近零售商店、公园的交通时长，并且使步行的路程更为切实可行。总之，丰富的道路系统支持了多种功能和多种建筑尺度。

LEARNING FROM JINAN: THE GRID CITY REVISITED

Jinan's Commercial North district is defined by its gridded road system that was built by the Germans in 1897. These major roads move traffic through the area but, in contrast with many newer Chinese developments, do so without creating impenetrable barriers that divide the neighborhood. In addition to these major through-streets, there are a variety of secondary streets and smaller paths that run within the interior of the blocks. These shortcuts reduce travel time to nearby retail, homes, and parks and make walking trips more feasible. Taken together, the range of road types support multiple functions and a variety of building scales.

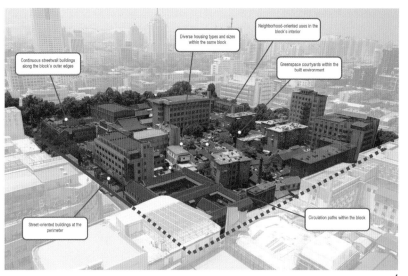

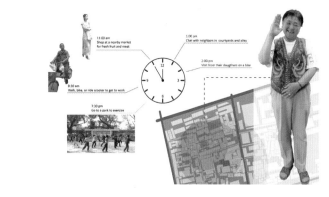

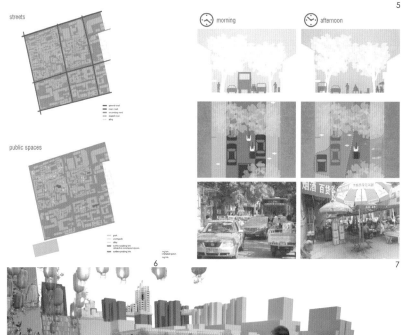

4 济南老城区示意 Jinan Old Town
5 在步行距离内混合用地 Mixed-Uses within Walking Distance
6 有层次的空间 Hierarchy of Spaces
7 灵活的街道 Flexible Streets
8 远景图 Future Perspective
9 总平面图 Master Plan
10 剖面图 Section

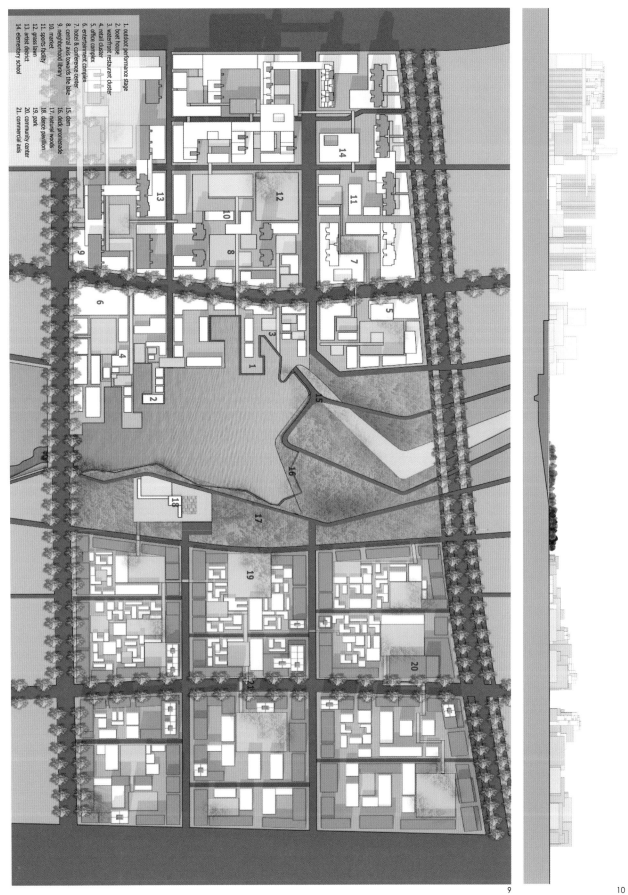

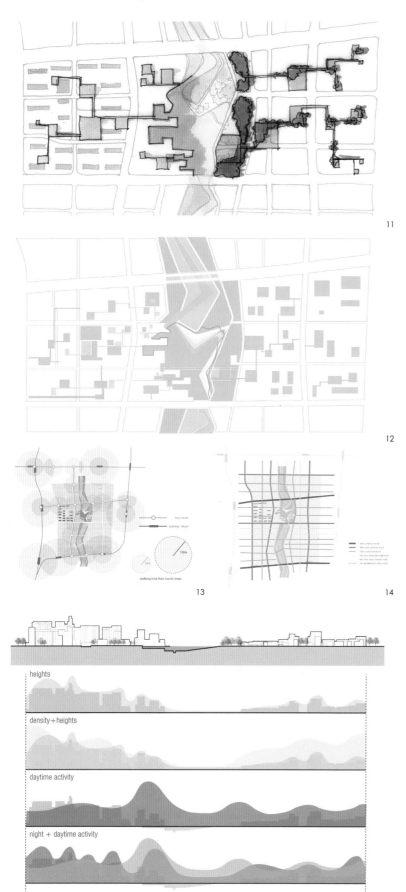

组织策略：有层次的场所

我们的场地坐落于北面的中央商务区和南面的大学城之间。尽管场地是混合使用的，但它以住宅区域的自然景观为特点，为这片发展中的新小区提供核心功能和社区环境。当工作日结束，人们会因它的社区化属性而聚集在此，而这一特征与北面商务区的非人性特征形成强烈对比。此外，周边丰富的街邻层次零售商店和集市为当地居民提供便利，让人们可以在这里实现绝大多数的日常生活需求，为减少交通能源的使用、培养邻里之间相互联系作出了贡献。

ORGANIZING METHOD: HIERARCHY OF PLACES

Our site sits between the CBD to the north and the university area to the south. Though the site will be mixed in use, it will be characterized by its residential nature, providing the heart and community of this new development district. When the workday finishes, this area is where people will congregate because it provides a community identity that contrasts with the somewhat impersonal nature of the commercial district to the north. Additionally, the neighborhood will provide a variety of neighborhood-level retail and markets that serve the population and allow people to carry out most of their daily functions within the site, contributing to lower transportation energy use and fostering interactions among residents living in the same community.

11 作为踏脚石的大小院落
Courtyards as Stepping Stones
12 院落层次
Hierarchy of Courtyards
13 路网层次 - 枢纽
Hierarchy of Transportation Network-Transit
14 路网层次 - 街道
Hierarchy of Transportation Network-Street
15 活力模拟
Vitality Simulation
16 改进现有住宅
Retrofitting Existing High-Rise Apartments
17 重新被诠释的路网城市
Reinterpreting the Grid City
18 充满活力的街区生活
Vibrant Streetspace

设计手法

我们计划将现状高层住宅和新的城市结构整合,进而将基地分为三个主要片区。在基地西部片区,现存建筑和不同的住宅类型融合,引进其他建筑来增强街道景观并增加地面和空中的联系。片区中心围绕一个湖面,成为基地的聚集点,西侧是一个服务周边社区的商业区,东侧是一个大型的自然景观公园,提供了一个重要的、适宜的开放空间。最后我们基地东部是全新建设的环境,遵循济南老商业中心的街区准则来运行:商业街类别的大型建筑物将街区联系起来,并在街区中心为不同住宅类型提供了一个空间。穿过我们的基地,庭院提供了有机的结构和关键的公共空间和设施。

DESIGN APPROACH

Our plan integrates the existing residential towers buildings into the new urban fabric and divides the site into three main areas. On the western portion of the site, the existing buildings are retrofitted with a variety of housing types, additional buildings are introduced to complete the streetscape and provide both ground level and elevated connections. The center of the site is organized around a lake that provides the focal point for the site. On the western side is a commercial district that serves the surrounding communities, and on the eastern side is a large, naturally landscaped park that provides an important open space amenity. Finally, on the eastern portion of our site is a completely new built environment that operates along many of the principles that organized Jinan's old commercial center's blocks: larger buildings with commercial street-level presence bound the blocks and provide an envelope for different housing types in the center of the block. Across our site, courtyards provide the organizational structure and critical public spaces and amenities.

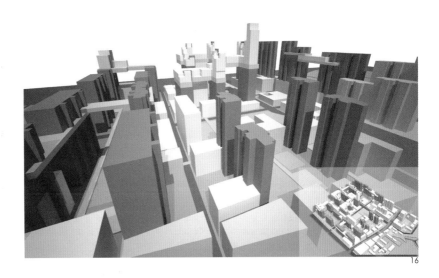

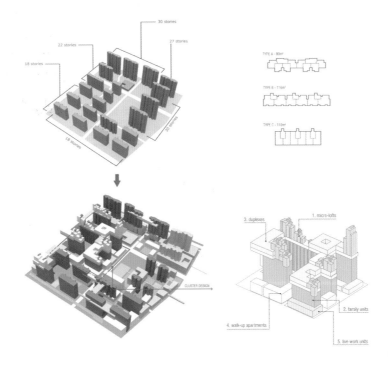

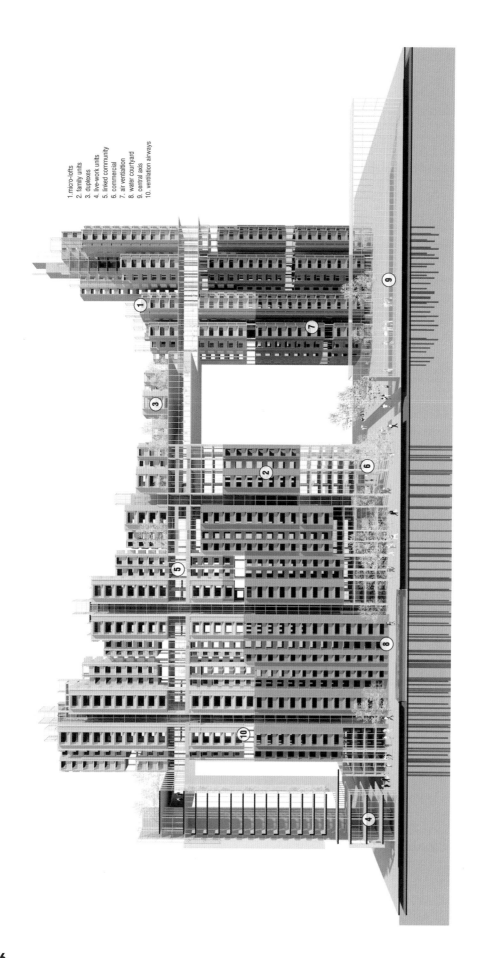

19 改进住宅示意
Retrofitting Aprtments

Existing

Retrofitted

micro-lofts

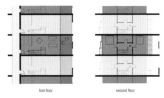
first floor / second floor
duplexes floor plan (scale 1:200)

live-work unit floor plan (scale 1:200)

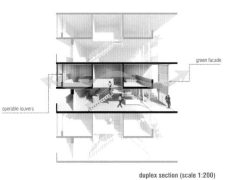
duplex section (scale 1:200)

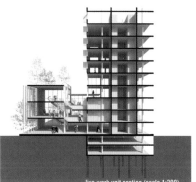
live-work unit section (scale 1:200)

20　　　　　　　　　　　　　21　　　　　　　　　　　　　22

23

20 户型设计 - 微户型 House Type-Micro Lofts
21 户型设计 - 复式户型 House Type-Duplexes
22 户型设计 - 住宅 - 工作室 House Type-Live-Work Units
23 透视图 Perspective

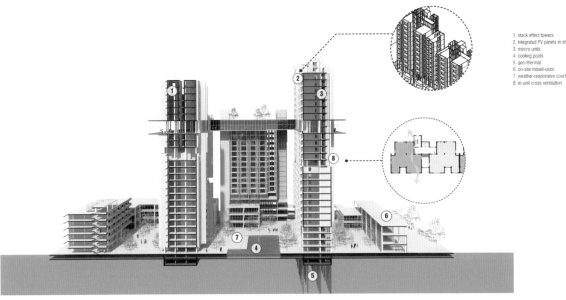

1. stack effect towers
2. integrated PV panels in shading elements
3. micro units
4. cooling pools
5. geo-thermal
6. on-site mixed-uses
7. weather-responsive courtyards
8. in-unit cross ventilation

25

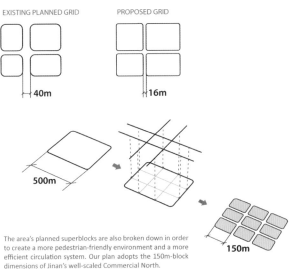

The area's planned superblocks are also broken down in order to create a more pedestrian-friendly environment and a more efficient circulation system. Our plan adopts the 150m-block dimensions of Jinan's well-scaled Commercial North.

26

Buildings at the edges of the blocks are oriented in a way that buffer the interior open spaces and buildings from the larger roads and higher levels of traffic. They also provide a continuous street wall that increases the quality of the pedestrian environment. The interior of the block is then left available for a variety of different kinds of housing types and community courtyard spaces. The highest buildings are located in the northeast corner of each block in a way that shields the rest of the block from the cold winter wind and makes the public open spaces useable for a greater proportion of the year. Lower buildings along the southern edges of the blocks let the southwest summer breeze cool the interior buildings and open spaces. By orienting the buildings in this way, our plan creates pleasant micro-climate and weather-responsive courtyards.

27

24 透视图
Perspective of Public Space
25 建筑级别策略
Architectural Level Strategy
26 邻里级别策略 - 打破超大街区
Neighbourhood Level Strategy- Break the Superblocks
27 邻里级别策略 - 微气候庭院
Neighbourhood Level Strategy- Respond to Environmental Systems
28 邻里级别策略 - 改善交通系统
Neighbourhood Level Strategy- Orient to Transit
29 邻里级别策略 - 混合用地
Neighbourhood Level Strategy- Mix Uses

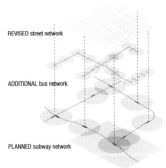

Existing transit plans for West Jinan include only subway and light rail systems that skirt the edges of our site. In order to create a robust transit-oriented development, our plan includes bus lines that will increase the accessibility to public transit and provide a needed hierarchy of transportation systems for different users, in much the same way that the road system does.

28

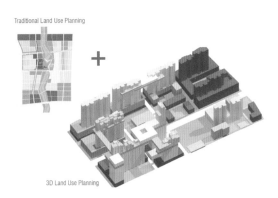

To reduce travel time, cost, and energy consumption and to create a truly vibrant neighborhood, mixed-uses are needed in not just the horizontal plane but also vertically.

29

Existing High-Rise Towers
Summer   Winter

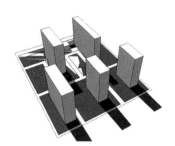
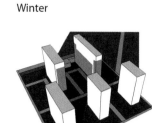

Retrofitted High-Rise Towers
Summer   Winter

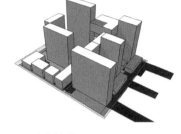

Reinterpreted Grid City
Summer   Winter

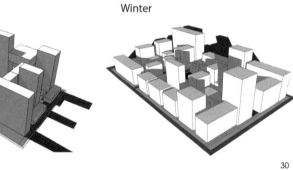

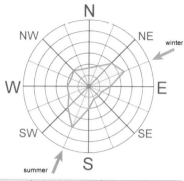

	Ave wind speed	Prevailing Direction
Winter	2.7m/s	ENE
Summer	2.8m/s	SSW

General Approach of the building layout is to block the praviling winter wind and open up for prevailing summer wind.

31

30 光照环境分析 Light Environment Analysis
31 风环境分析 Wind Environment Analysis
32 能量性能分析 Energy Proforma Analysis

30

Tower in the park	Retrofitted towers	Reinterpreted Grid City
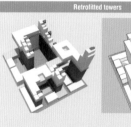		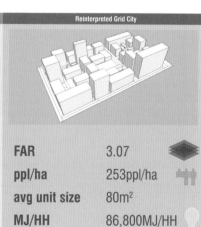
FAR 1.76	FAR 5.30	FAR 3.07
ppl/ha 130ppl/ha	ppl/ha 415ppl/ha	ppl/ha 253ppl/ha
avg unit size 97m²	avg unit size 80m²	avg unit size 80m²
MJ/HH 101,400MJ/HH	MJ/HH 73,500MJ/HH	MJ/HH 86,800MJ/HH

32

活力新城
VILLAGE FOR VITALITY
CAINE, HOWLAND, LIN, LIU, WENG, YU

藉由连结个人、团体及小区生态环境的发展策略，本案「活力新城」旨在兼顾济南西部新城开发与自然环境的保护。又本案是美国麻省理工学院与中国清华大学联合都市计划课程的实习作业，该课程主要目的在于探讨减少都市能耗、可持续发展及智能型都市成长。城市的生产效能源自于城市环境的健康程度，本设计的目的是提升空气质量、降低能耗、鼓励绿色建筑、创造多样绿地、绿色交通（电力车及共享脚踏车措施）及资源回收。「活力新城」的规划成果展现本组对于可居住性城市的想法，藉由本案的都市设计手法，希望可以提供济南市民更健康及舒适的都市生活环境。

Village for Vitality (VV) is dedicated to increasing the compatibility of West Jinan with the natural environment by providing sustainable development strategies to link individuals, groups and community-based ecological activities. VV is a project of the MIT-Tsinghua Joint Urban Design Studio that promotes the concept of energy efficiency and emphasizes sustainability with smart urban growth. The goal of our project is to improve the quality of the air, lower the use of non-renewable resources, encourage the building of green homes, offices, and other structures, reserve more user friendly green space, support environmentallyfriendly methods of transportation (electric vehicle and bicycle sharing program), and offer recycling programs. VV embraces livable-city principles and design strategies that enhance the health and wellbeing of citizens in urban environments.

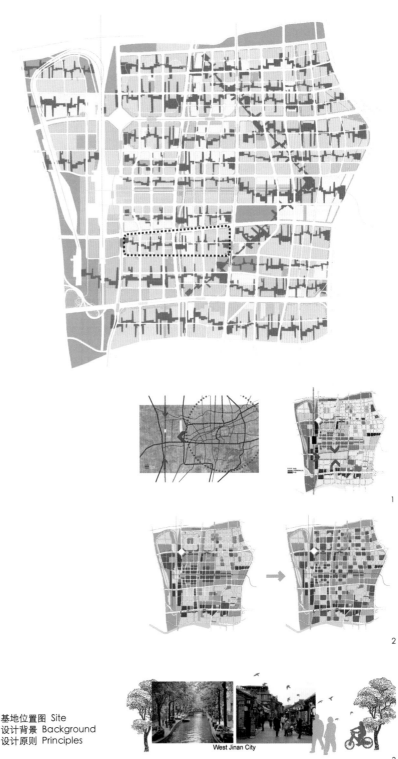

1 基地位置图 Site
2 设计背景 Background
3 设计原则 Principles

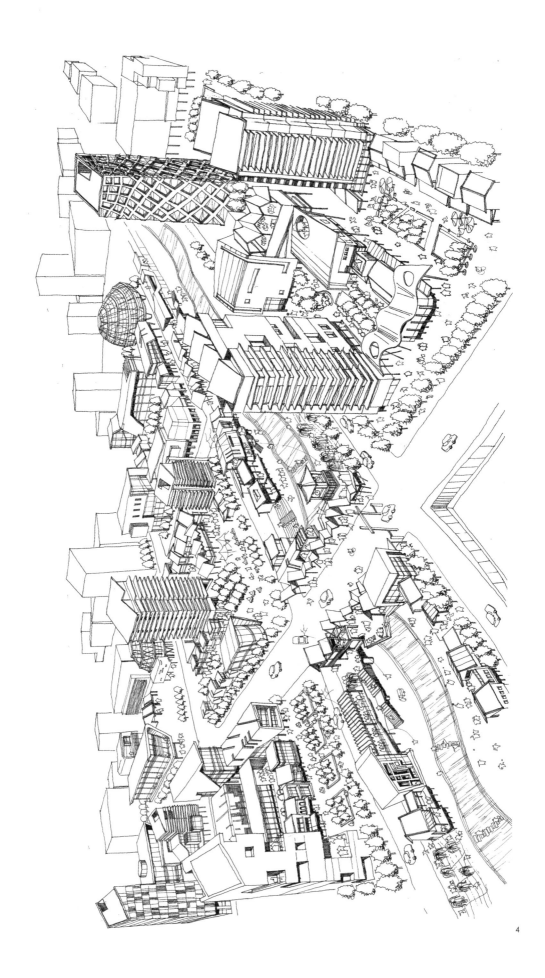

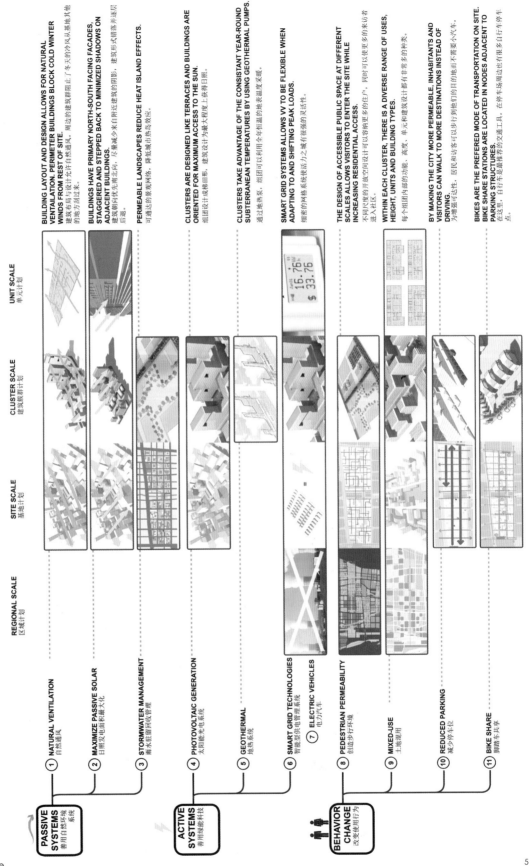

4 透视图 Perspective
5 节能设计原则 Principles of Energy Conservation Design

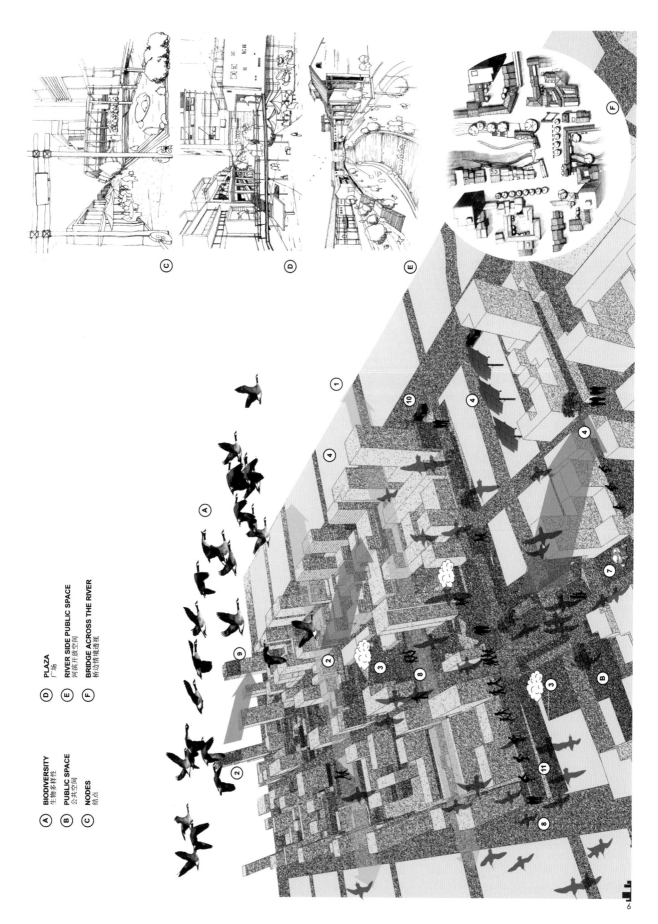

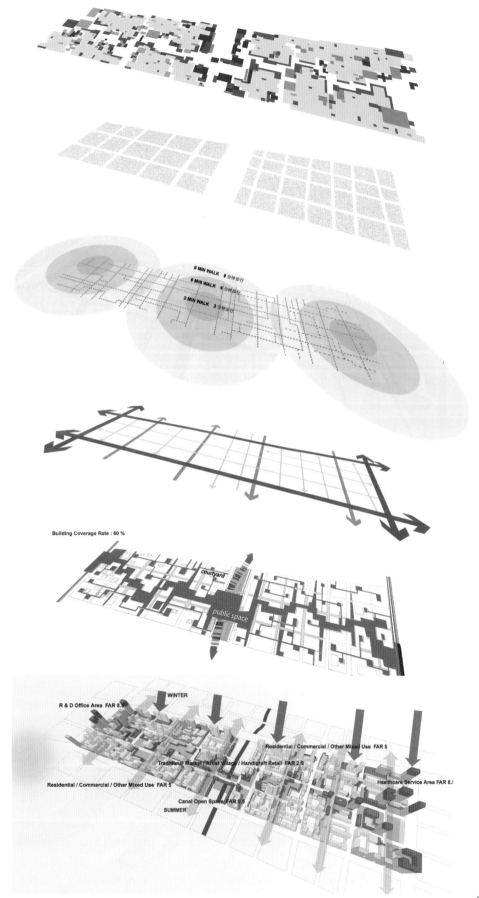

6 重要节点及透视
Main Nodes and Perspective
7 基地整体设计策略
Integrated Design Strategies

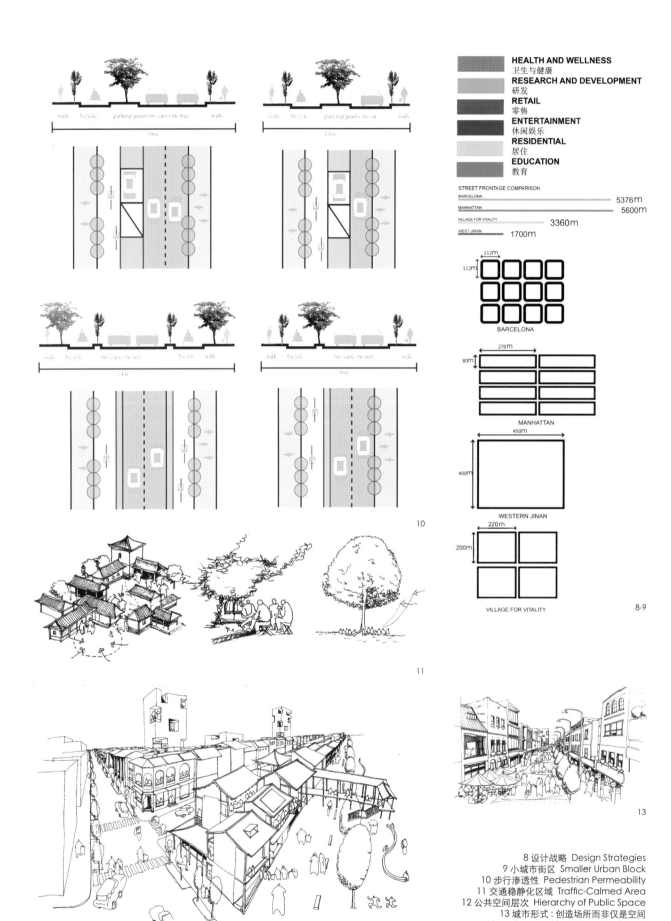

8 设计战略 Design Strategies
9 小城市街区 Smaller Urban Block
10 步行渗透性 Pedestrian Permeability
11 交通稳静化区域 Traffic-Calmed Area
12 公共空间层次 Hierarchy of Public Space
13 城市形式：创造场所而非仅是空间
Urban Form:Making Place Not Space

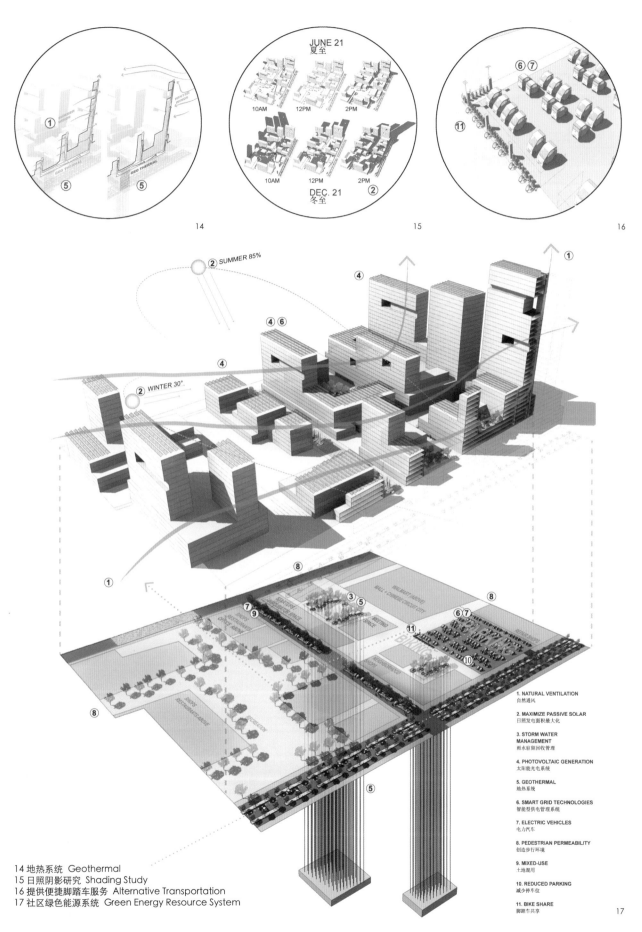

14 地热系统 Geothermal
15 日照阴影研究 Shading Study
16 提供便捷脚踏车服务 Alternative Transportation
17 社区绿色能源系统 Green Energy Resource System

2 BDRM = 105 SM 1 BDRM = 59 SM ← UPGRADE PLAN 2

2 BDRM = 96 SM 1 BDRM = 59 SM ← UPGRADE PLAN 1

2 BDRM = 92 SM 1 BDRM = 51 SM ← BASELINE PLAN

JUNE 21
2PM

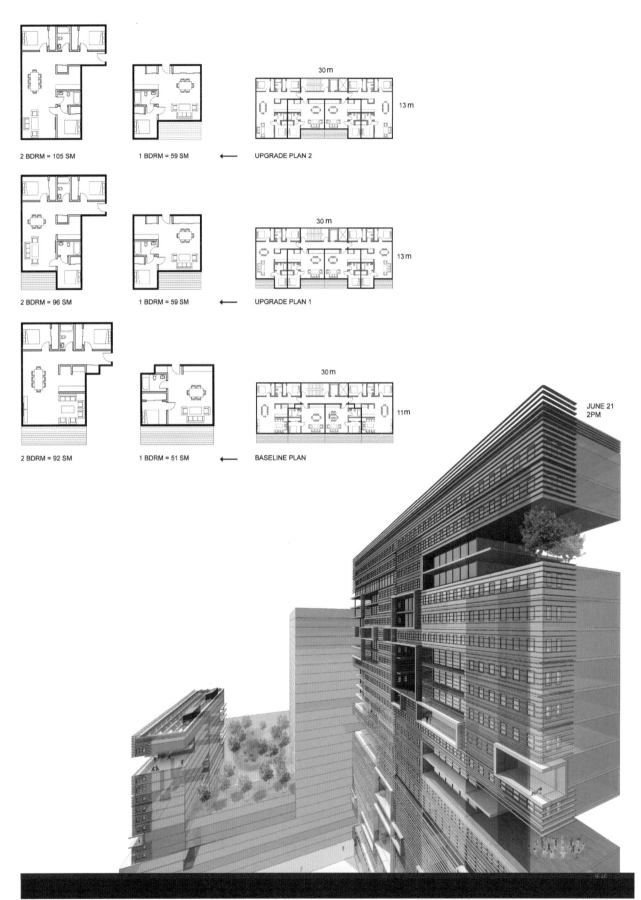

| FAR 容积率 | 3.09 |
| DENSITY 密度 | 24,400 HH/KM² |

ENERGY CONSUMPTION (MJ/HH)
能源排放
OPERATIONAL 运营能耗	50.6
TRANSPORTATION 交通能耗	12.9
EMBODIED 内部能耗	22.8
RENEWABLE 更新能耗	5.05
TOTAL 总计	81.1

| FAR | 3.78 |
| DENSITY | 26,543 HH/KM² |

ENERGY (MJ/HH)
OPERATIONAL	52.1
TRANSPORTATION	13.7
EMBODIED	26.6
RENEWABLE	5.4
TOTAL	87

| FAR | 2.4 |
| DENSITY | 22,254 HH/KM² |

ENERGY (MJ/HH)
OPERATIONAL	49
TRANSPORTATION	12
EMBODIED	18.9
RENEWABLE	4.7
TOTAL	75.2

| FAR | 2.15 |
| DENSITY | 16,624 HH/KM² |

ENERGY (MJ/HH)
OPERATIONAL	72.4
TRANSPORTATION	14.8
EMBODIED	29.8
RENEWABLE	0
TOTAL	117

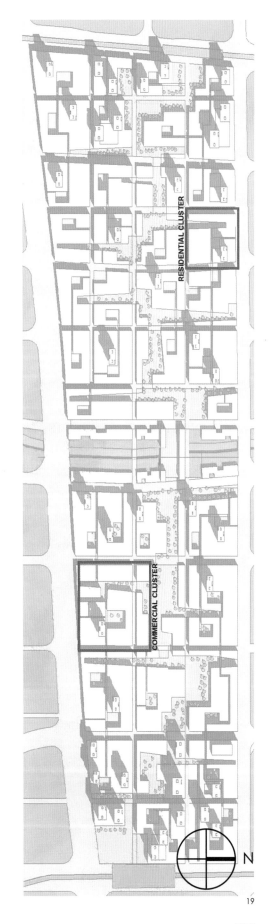

18 户型设计 House Types
19 邻里的能耗 Neighbourhood Performance

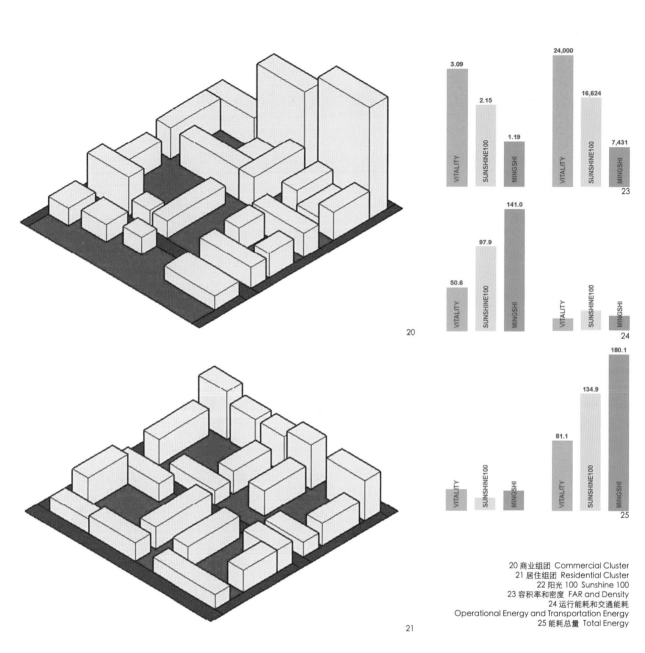

20 商业组团 Commercial Cluster
21 居住组团 Residential Cluster
22 阳光 100 Sunshine 100
23 容积率和密度 FAR and Density
24 运行能耗和交通能耗
Operational Energy and Transportation Energy
25 能耗总量 Total Energy

活力之城的公式数据与从阳光100小区以及名仕地产得来的实证调查数据相比较，活力之城比这两者达到了更高的容积率和密度，活力之城的预计运营能耗也明显更低。

Village for vitality PR of ORMA data is compared to empirical survey data from Sunshine 100 and Mingshi Developments. Village for vitality achieves a higher far and density than either Sunshine or Mingshi. Village for vitality's predicted operational energy is significantly lower as well.

济南西部新城
OASIS NEIGHBORHOOD

Lun Lui, Slobodan Radoban, Gilad Rosenzweig, Alice Shay, Wang Zhe

1 总平面图 Master Plan
2 济南 Jinan
3 新邻里类型 New Neighbourhood Type
4 连续的景观 Continuous Landscape
5 院落结构 Courtyard Structure
6 共享空间单元系统 Shared Space Unit System
7 院落案例：柏林和温哥华
Courtyard Precedents: Berlin & Vancouver

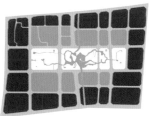

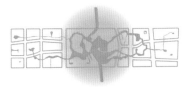

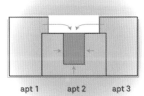

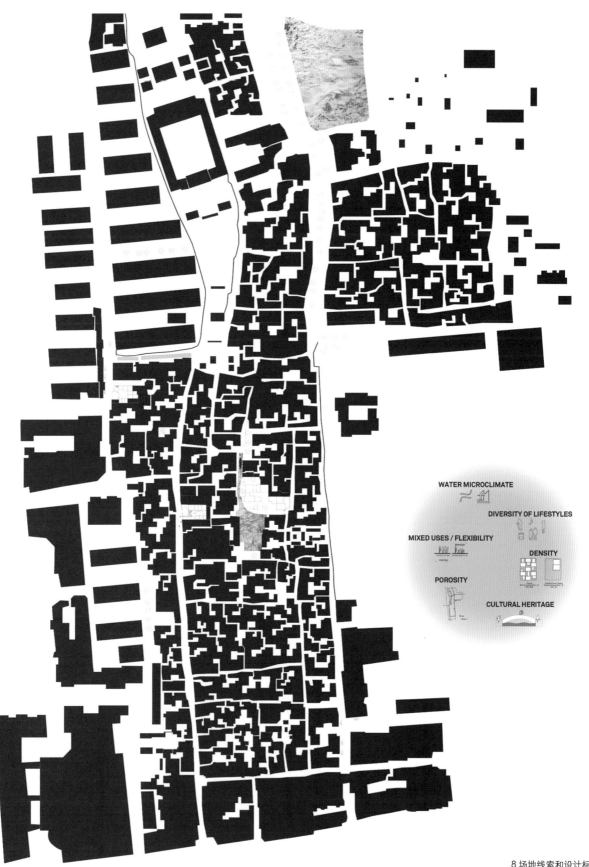

8 场地线索和设计标准
Site Clues and Design Criteria
9 不同使用者分析
Analysis of Different Users

BUS DRIVER:
CIRCULATION AND STREET TEXTURE

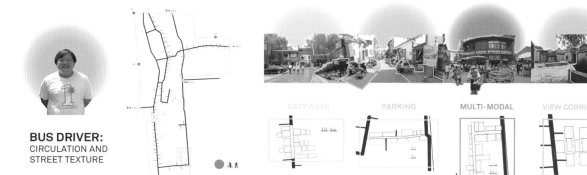

GATEWAYS PARKING MULTI-MODAL VIEW CORRIDORS

Residents choose cheap energy (gas vs. electricity).

SNACK SHOP OWNER:
COMMERCIAL AND WORK PATTERNS

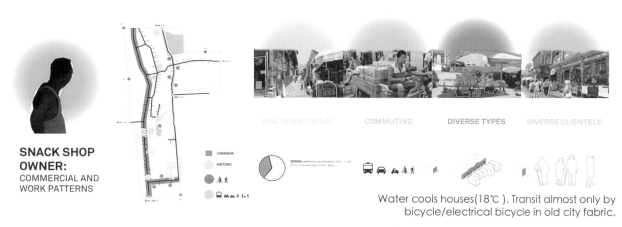

MULTIFUNCTIONAL COMMUTING DIVERSE TYPES DIVERSE CLIENTELE

Water cools houses (18°C). Transit almost only by bicycle/electrical bicycle in old city fabric.

RESTAURANT OWNER:
HOUSING AND LIVING PATTERNS

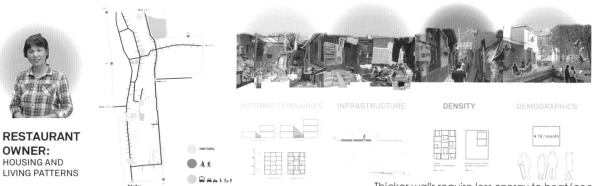

HISTORIC TYPOLOGIES INFRASTRUCTURE DENSITY DEMOGRAPHICS

Thicker walls require less energy to heat/cool. Well-shaded rooms, balconies.

INSTRUMENT CRAFTSMAN:
PUBLIC REALM AND WATER LANDSCAPE

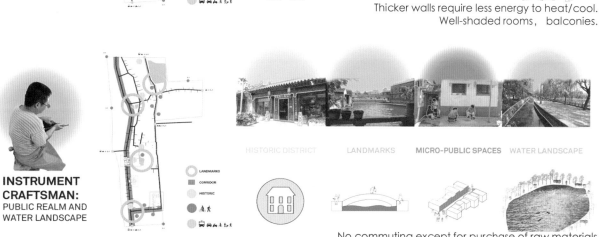

HISTORIC DISTRICT LANDMARKS MICRO-PUBLIC SPACES WATER LANDSCAPE

No commuting except for purchase of raw materials

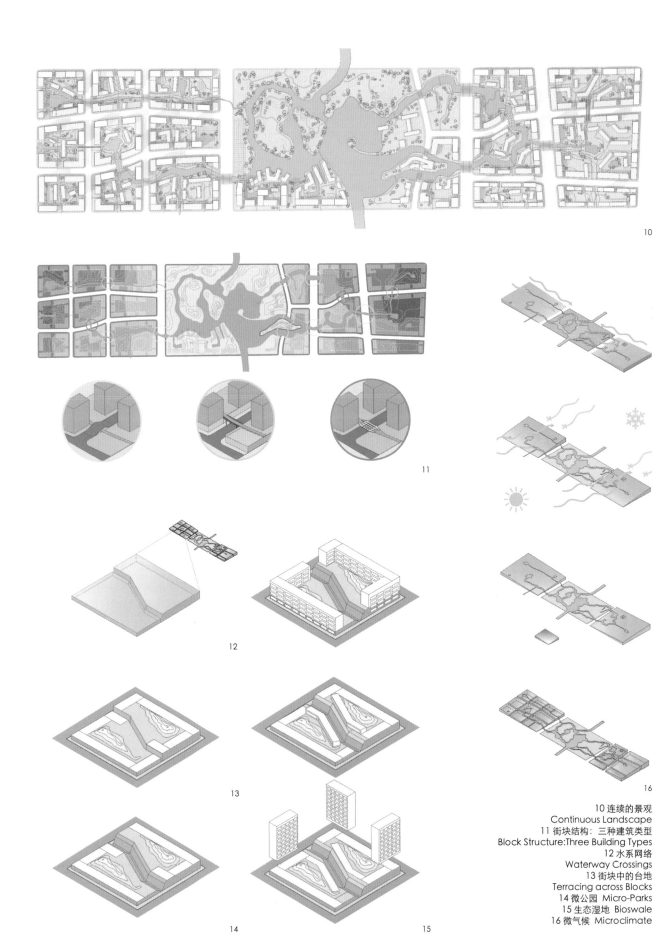

10 连续的景观 Continuous Landscape
11 街块结构：三种建筑类型 Block Structure: Three Building Types
12 水系网络 Waterway Crossings
13 街块中的台地 Terracing across Blocks
14 微公园 Micro-Parks
15 生态湿地 Bioswale
16 微气候 Microclimate

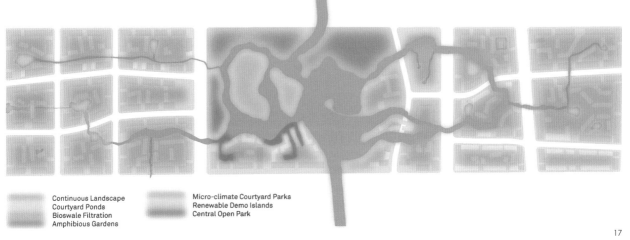

Continuous Landscape
Courtyard Ponds
Bioswale Filtration
Amphibious Gardens

Micro-climate Courtyard Parks
Renewable Demo Islands
Central Open Park

17

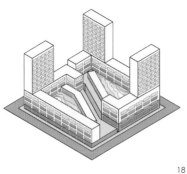

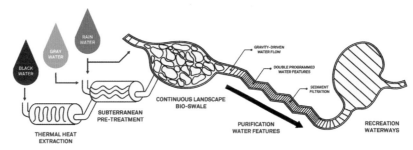

18

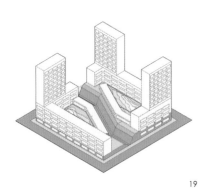

19

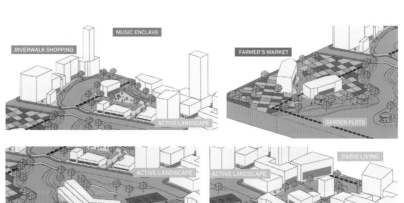

20

17 活跃的景观
Active & Productive Landscapes
18 透气的院落 Porous Courtyards
19 湿／干水系廊道
Wet/Dry Aquatic Dorridor
20 中心王国 Middle Kingdom

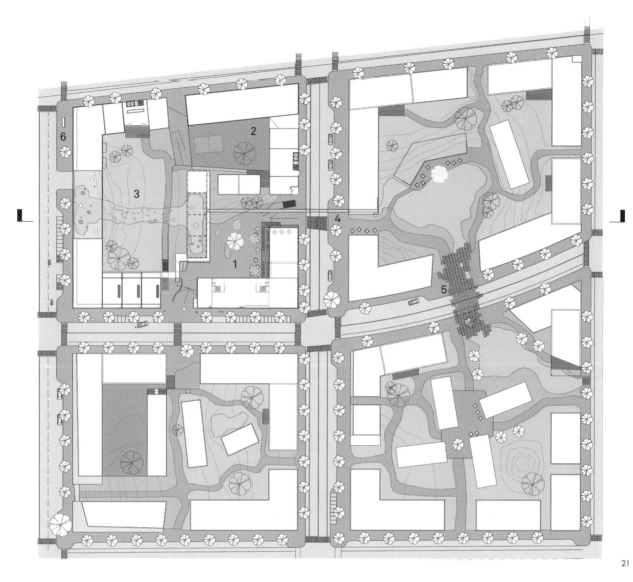

Exemplary Uses

1. Public Square 公共广场
2. Private Courtyard 私人院落
3. Bioswale 生态湿地公园
4. Aqueduct 沟渠
5. At grade water crossing 梯度水系
6. Local bus stop 公交车站
7. Electric "City" car station 电动小车站
8. Bicycle parking 自行车泊车
9. Water reclamation 水回收

21 1:500 街区平面图 Cluster Plan_1:500
22 剖面图 Section

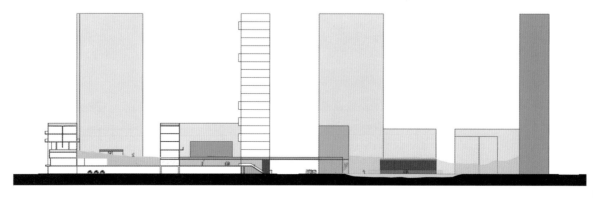

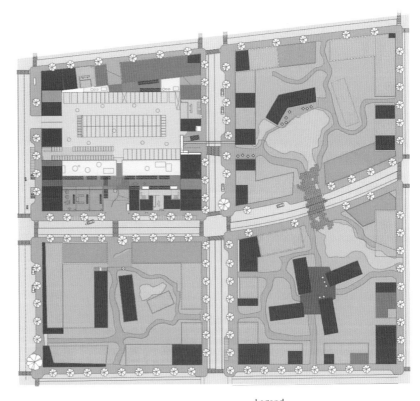

23 生态湿地 Bioswale
24 公共广场 Public Courtyard
25 土地利用图 Land Use Plan
26 首层平面图 Ground Floor Plan

Legend

Residential	零售
Commercial	商业
Institutional	公共服务设施
Live/Work (s.o.h.o)	居住/工作

25

23

24

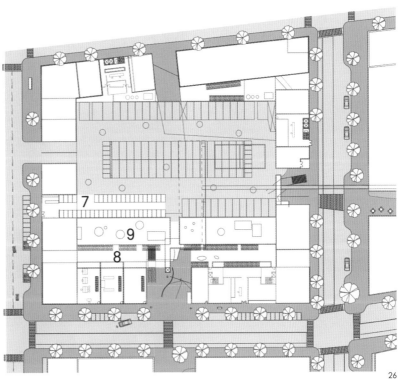

26

117

● Age ● Family Size ● Hobby ● Activities

Sleeping | Hygiene | Working | Socializing | Eating | Cooking | Washing | Storage

Bedroom | Bathroom | Office | Living Room | Dining Room | Kitchen | Laundry Room | Storage Room

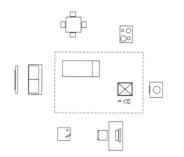

27 生活方式多样性
Diversity of Lifestyle
28 日常活动的空间组织
Spatial Organization of Daily Activities
29 芙蓉街的院落活动
Courtyard Activities in Furong
30 邻里活动
Neighborhood Activities
31 共享空间 vs. 非共享空间
Shareable vs. Non-shareable Spaces
32 作为居民联系的芙蓉院
Furong Courtyard as Connector Between Residents

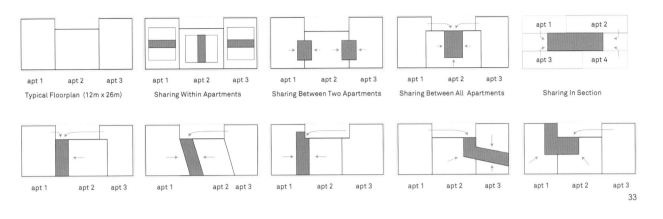

33

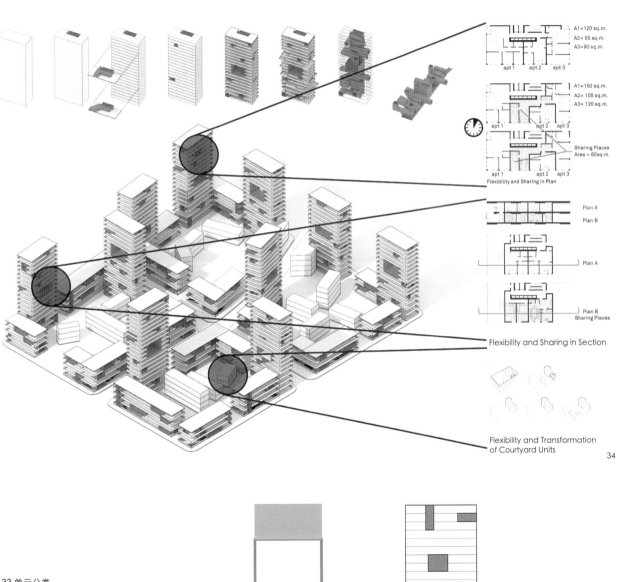

34

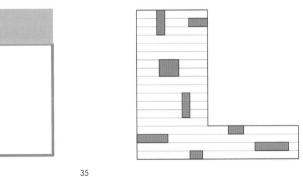

33 单元分类
Unit Taxonomies
34 单元设计法则
Unit Design Principles
35 底层面积增加
Gain in Floor Area
36 社区活动的强调
Emphasis of Community Activities

35   36

Annual Energy Consumption (MJ per HH)

Energy Proforma Metrics	
Building height	26 m
Building coverage	28 %
South-facing wall ratio	28 %
Surface-to-volume ratio	26 %
Bldg. Function Mix	12 %
Avg. Bldg Footprint	523 sqm
Percentage GF Comm	35 %
Roof PV %	33 %
Avg Unit Size	80 sqm
FAR	2.7
Population Density	245 HH / Ha

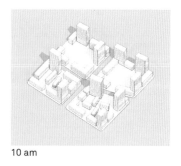
10 am

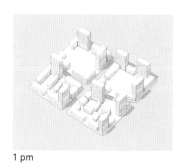
1 pm

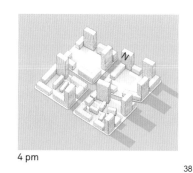
4 pm

1. Building form and position respond to air flow.
2. Water recycling: black, gray and rain water.
3. Landscape allows for underground activities to take place on earth.
4. Curbing an urban heat island.
5. Trash collection withing each floor.
6. Sharing daily activities.
7. Photovoltaic solar energy.
8. Natural light in the core of the building.
9. Car and Bicycle sharing.
10. Urban Porosity / Proximity / Neighborhood amenities.

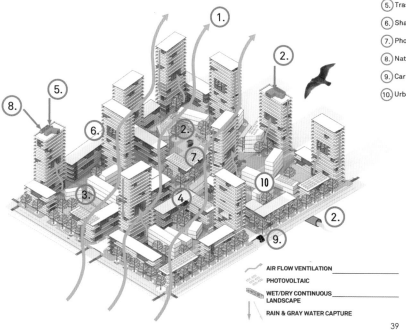

AIR FLOW VENTILATION
PHOTOVOLTAIC
WET/DRY CONTINUOUS LANDSCAPE
RAIN & GRAY WATER CAPTURE

37 济南新城组团模型的结果
Proforma Results based on massing model of clusters of the Jinan Oasis
38 阴影分析 Shadow Analysis
39 社区整体绿色能源系统
Whole Green Energy System in Neighbourhood

生产的城市
THE PRODUCTIVE CITY
Elliott-Ortege, Ethan Lay-Sleeper, Boris Berndtson, Kun Qian, Jixiao Wang

1 2 3

生产性城市规划设计是对城市人口、工业和生活方式的综合应对，在时间维度上考虑混合和弹性元素的调整来满足社区居民的需求。

The Productive City is one that can respond to changes in population, industry, and lifestyle. Fixed and flexible elements are adapted over time to suit the needs of neighborhood residents.

4

1 适应性框架 Adaptable Framework
2 分层混合 Layered Programming
3 综合生产 Integrated Production
4 邻里平面 Neighbourhood Plan

production　tech startups　markets　topography　education

121

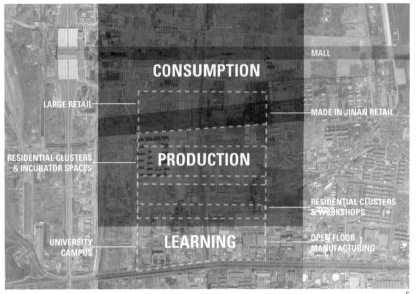

轨道交通站点是社区的自然入口，五分钟步行可达距离全范围覆盖。除此之外，两条商业街从南部道路的地铁站点向北延伸。社区内共有4个市场，每个都有其独有的特性，例如，一个是专门为济南本地特产设置的展示平台，另一个主要服务于周边的大学生。

The metro stations are the natural entry points to the neighborhood; five-minute walking radii are displayed above. Additionally, two retail corridors extend northward from the stations along the southern road. Four markets are supported in the neighborhood, each with its own identity. One showcases Made-in-Jinan products, for intance, while another is primarily aimed to serve university students.

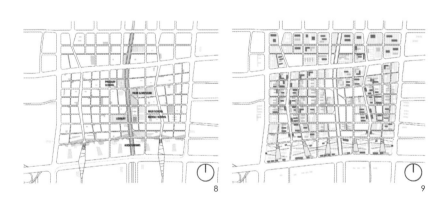

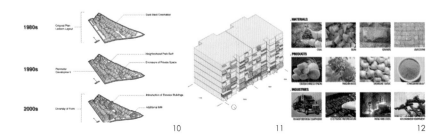

5 生产的机制 The Case for Production
6 济南的西部门户 Gates to Jinan West
7 社区市场 Neighbourhood Markets
8 内嵌的公共设施 Nested Public Facilities
9 多层次的项目 Layered Programs
10 生产结构体系 Enclave Framework
11 严谨的形式 Articulated Form
12 本地资源 Local Resources

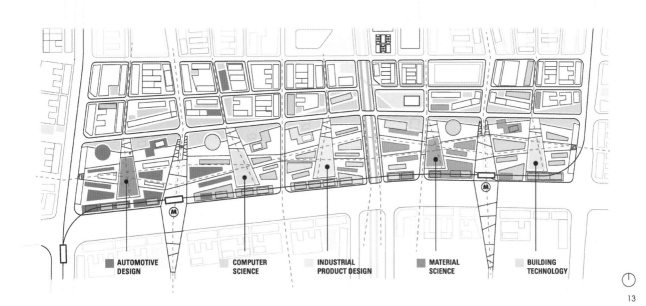

济南西部大学是突破传统形式和功能的未来校园，作为一个教授各类职业技术的综合类院校，每个学院都有一个专门的大型实践场地，在此之上再建设主要建筑，这些建筑通过东西人行走道互相连接，同时也促进了各学科之间的交流。

Jinan West University is the university of the future, breaking away from the form and function of the traditional campus. As a polytechnic institution focused on applied skills, each college features a large manufacturing space, above which is located its main quad. The quads are connected via an east-west pedestrian band that facilitates interdisciplinary movement.

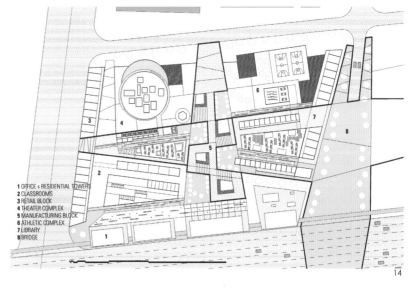

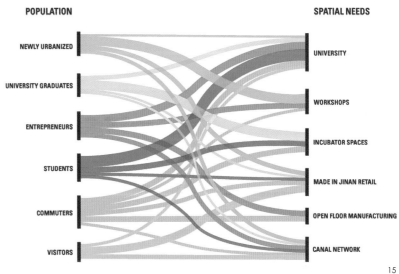

13 地段平面 Area Plan
14 济南西大分层混合
The Colleges of Jinan West University
15 综合生产
Overlapping People and Places

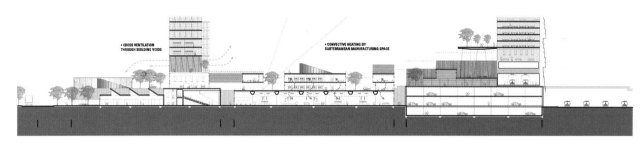
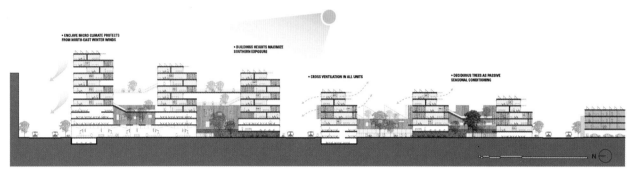
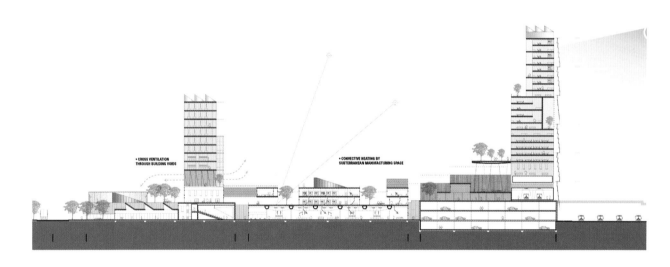
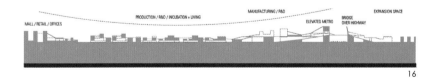
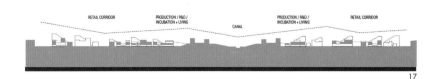

16 东西方向剖面 East-West Section
17 南北方向剖面 North-South Section

参考既有成功生产模式，加入三维空间的连接，这种多层次的综合生产模式不仅提供了一种独特的生产性公共空间结构，还为不同居住单元提供了交流平台。

Taking the already successful enclave form and adding three-dimensional articulation, the multilayered enclave provides a unique collection of productive and public spaces, in addition to a platform for a diversity of residential units.

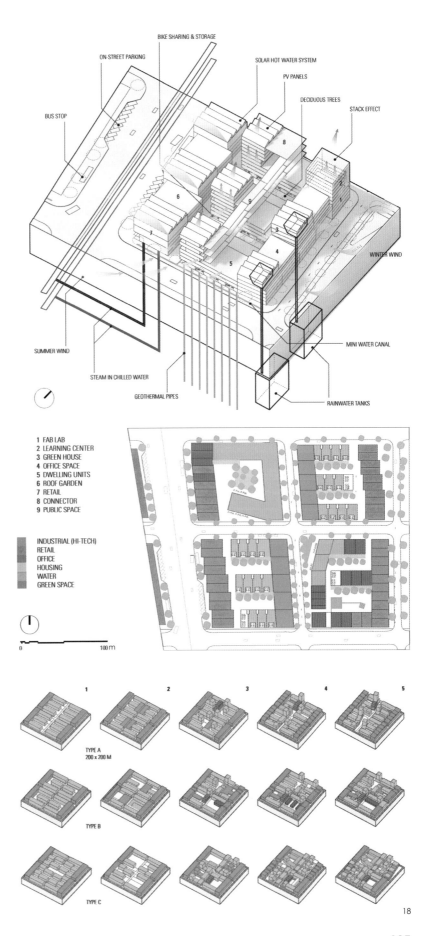

18 综合生产 Multilayered Enclave

济南西部大学作为社区的入口，并且由于其毗邻地铁站点，它也是一个交流汇集的场所，包括大型广场及其上的购物中心等可达的公共空间。光线具有良好的将公共视线引入下方活动场所的作用，由于大学很好地融入了城市结构体系中，其作为社区的中转站几乎是无缝衔接。

West Jinan University serves as the gateway to the neighborhood, as it is adjacent to the metro stations. It is also a community space, with publicly accessible spaces including the large plazas atop manufacturing spaces. Light wells invite the public to view the activities taking place below. The transition to the neighborhood is relatively seamless, as the university is well-integrated into the urban fabric.

19 多角色校园
On/Off Campus
20 吸纳企业进驻
Enabling Entrepreneurship
21 产业集聚效应
Industrial Agglomeration Effect
22 空间单元及结构组织
Unit and Framwork

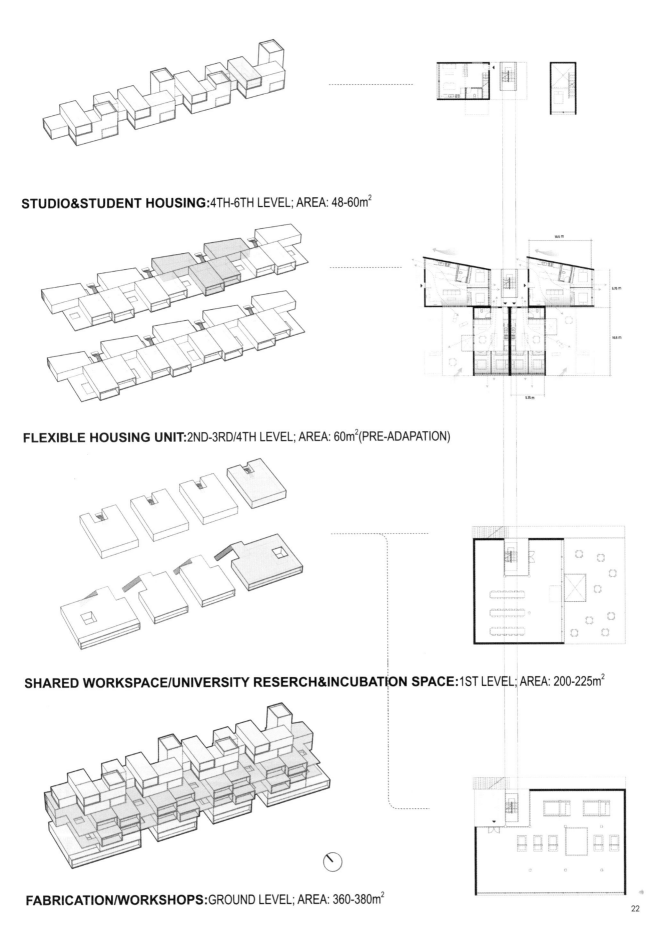

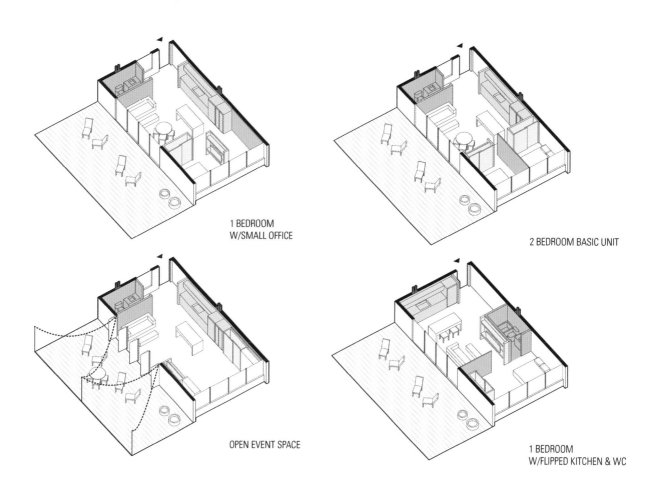

FLEXIBILITY THROUGH PREFAB COMPONENTS

- PLUMBING CORE / PLUG IN POINTS
- BATHROOM & KITCHEN UNIT
- PARTITION & STORAGE UNIT

3 - 2BED UNITS
60 m²

1 - 1BED + OFFICE
60 m²

1 - STUDIO,
55 m²

1 - 2BED
60 m²

1 - 4BED
130 m²

2 - 2-BED
60 m²

1 - 2BED + OFFICE 1 - EFFICIENCY
85 m²

+
1 - EFFICIENCY UNIT
35 m²

2 - 2BED
60 m²

1 - 3BED
70 m²

1 - 1BED
45 m²

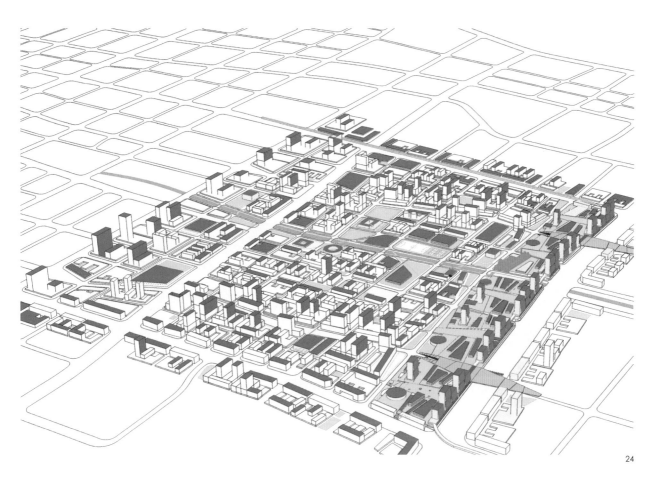

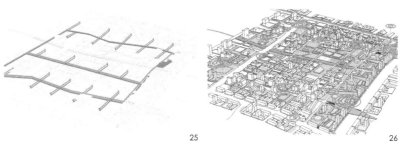

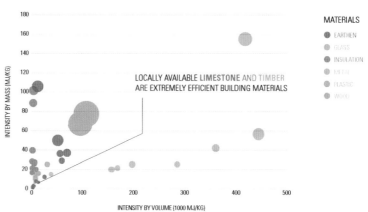

23 户型分析 House Types
24 太阳光电系统 Solar Photovoltaics
25 分散式发电布局 Distributed Generation
26 共享的移动网络 Shared Mobility
27 本地材料 Local Materials

利用沿着社区南部边界的两个地铁站点，"生产的城市"是紧凑的、适宜步行的、交通主导的。太阳能发电系统在社区内全覆盖布局，另外，在这个综合性大学内部还引入了一个沼气废热发电厂。经过这些灵活可变空间及其适宜发展尺度的精心考虑，没有牺牲任何居住容量或降低能源效率就实现了高密度。

Taking advantage of the two metro stations along the southern edge of the neighborhood, the Productive City is compact, walkable, and transit-oriented. Solar energy generation is sited throughout the neighborhood, and a biogas cogeneration plant is introduced at the polytechnic university. Through the careful consideration of adaptable space and right sizing, high density is achieved without sacrificing livability or energy efficiency.

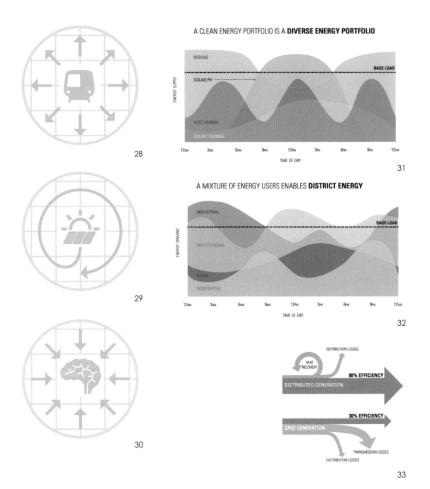

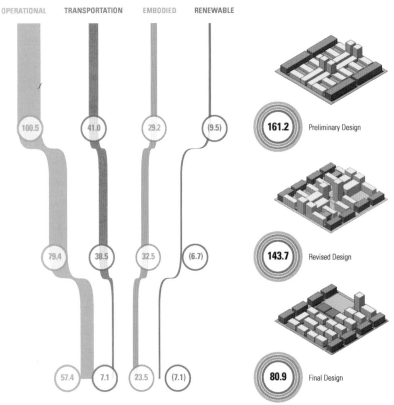

28 紧凑发展 Compact Development
29 公共产能 Public Generators
30 精明增长 Smart Density
31 动态供给 Dynamic Supply
32 本地效益 Local Efficiency
33 平衡需求 Balanced Demand
34 自循环 Proforma Iterations

2014

太原　　TAIYUAN
城中村　URBAN VILLAGE
清洁能源城市　CLEAN ENERGY CITIES

城市更新中如何体现可持续发展的策略，降低碳排放，是与城市新区开发同等重要的议题。本次联合设计选择了山西太原作为研究对象，对城市存量土地中的既有居住区、城中村的更新展开研究，探讨可持续的城市设计手法在更新改造中的应用。地段涉及复杂的社会结构和多样的居住形态，和城市的建成区已经建立了错综复杂的功能关系，如何在更新过程中解决建筑与空间的更新，同时系统考虑已经存在的社会、文化关系，并实现降低碳排放、提升经济活力的目标，是此次规划设计研究的重点。

How to apply the sustainable development to urban renewal strategy and reducing carbon emissions, are the key issues equaling to city new district development. The joint urban design studio chose Taiyuan in Shanxi Province as the research object, studying the existing residential area and the updating of urban villages in the urban stock land, to explore sustainable urban design methods in the application of the renewal and reconstruction. Complicated social structures and multiple residential typologies are involved in this site, which built complicated functional relations to the existing urban area. The question that how could existing social, cultural relations cooperate with the architectural and spatial retrofitting process, and how to achieve the goal of lower carbon commission, higher economic activeness is what we have concerned most during this planning.

五种
FIVE SEEDS

Kara Elliott-Ortege, Ethan Lay-Sleeper, Boris Berndtson, Kun Qian, Jixiao Wang

1 设计场地 Site
2 现场考察 Local Experience

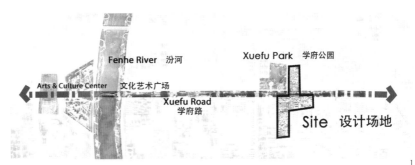

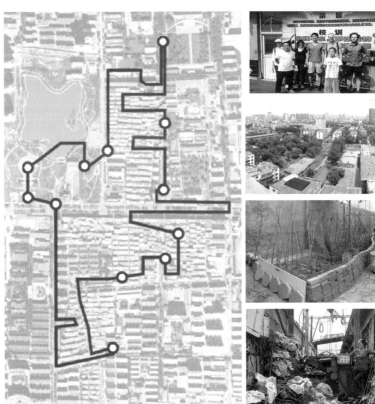

可持续发展的城市设计不仅仅要求节约能源，更重要的是为城市源源不断地注入活力。依托地段内已有的功能活动，结合城市区位和周边条件，"五种"通过植入五个城市功能组团，以带动城中村的产业，引发其自我更新与功能系统完善，最终成为自然生态与社会生态综合协调的有机聚落。

A sustainable urban design not only requires energy saving, it is more important to constantly inject energy into the city. Implementing 5 programmed clusters, FIVE SEEDS intends to stimulate the urban village renewal through its local industry.

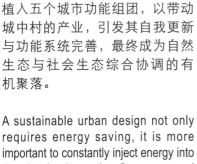

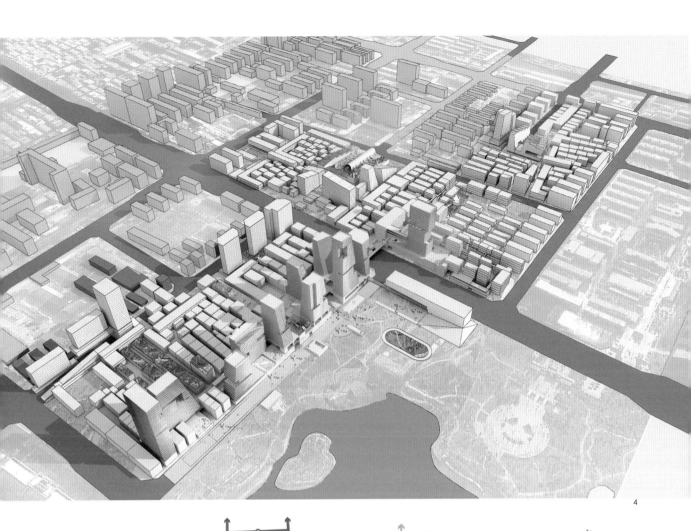

4

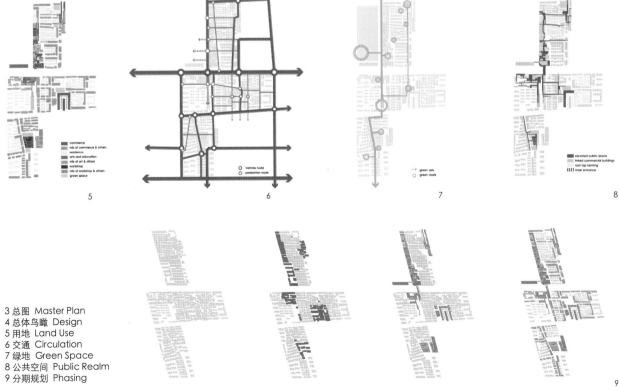

5　　　　　　　　　　6　　　　　　　　　　7　　　　　　　　　　8

9

3 总图　Master Plan
4 总体鸟瞰　Design
5 用地　Land Use
6 交通　Circulation
7 绿地　Green Space
8 公共空间　Public Realm
9 分期规划　Phasing

135

10

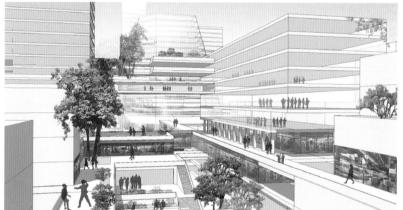

11

12

10 新边界 - 艺术文化 New Edge-Art&Culture
11 艺术培训中心 Art Education Center
12 学府公园东望 View From the Park
13 分层研究 Layers

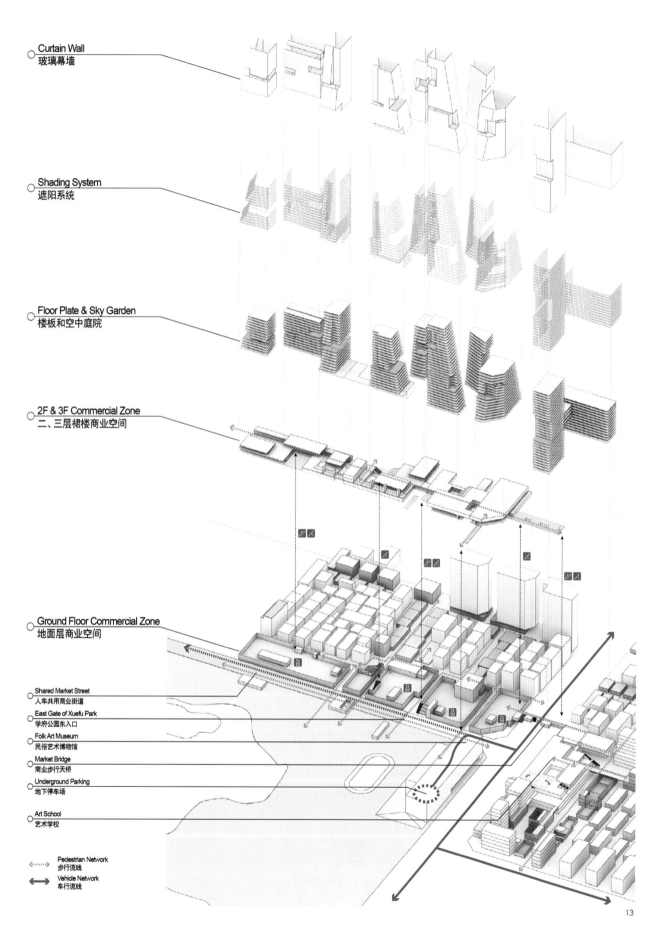

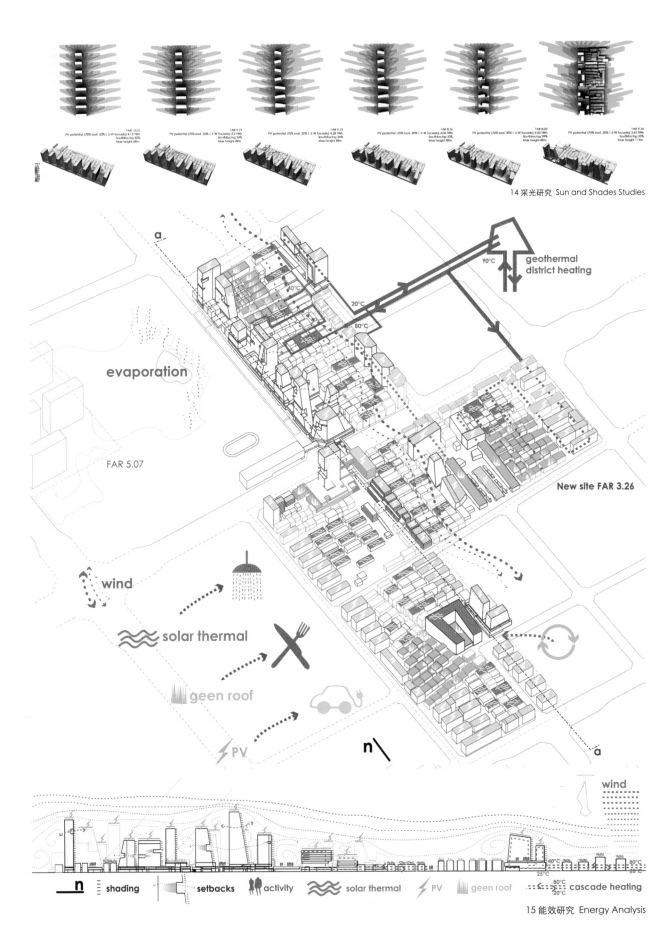

14 采光研究 Sun and Shades Studies

15 能效研究 Energy Analysis

社会形式
SOCIAL PROFORMA
Agustina, Paloma, LongRei, Jan Bo

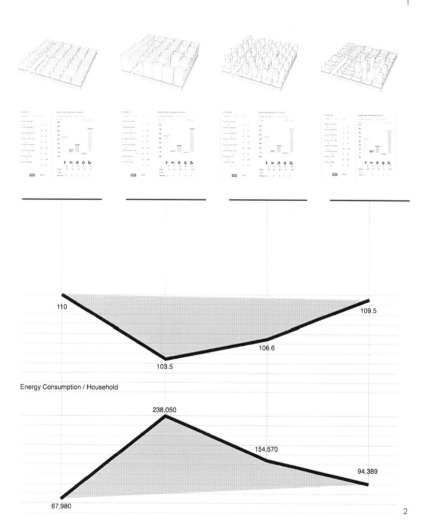

1 场地现状记录 Field Research
2 现有住房能耗构成
Energy Consumption of Housing

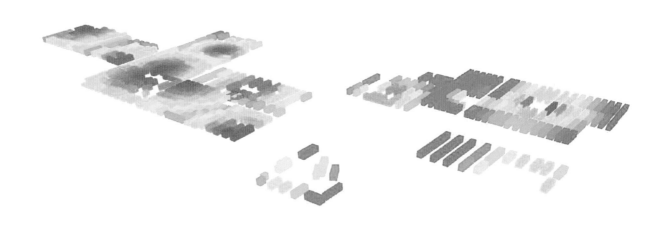

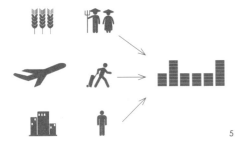

3 重新划定设计范围 Design Area
4 用地范围内公共性程度 Publicity in the area
5 设计策略 Design Strategy
6 设计总图 Master Plan

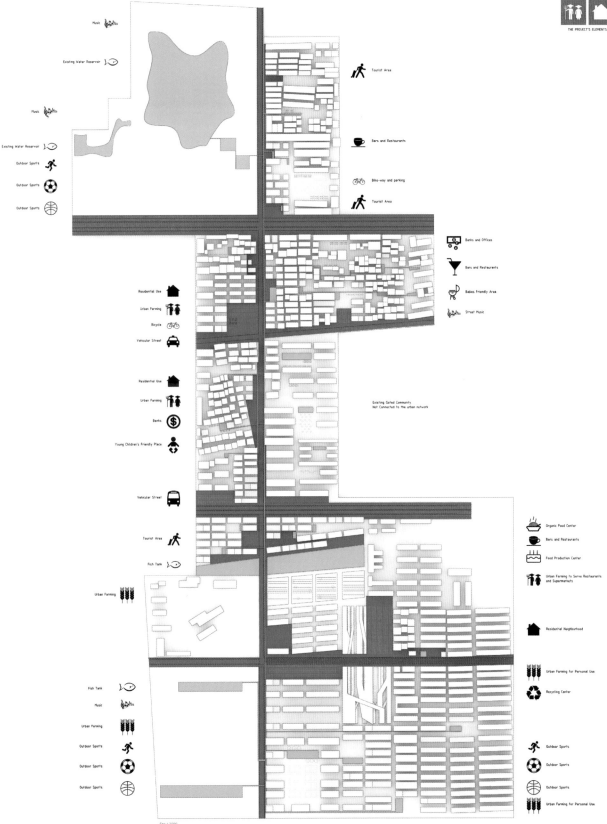

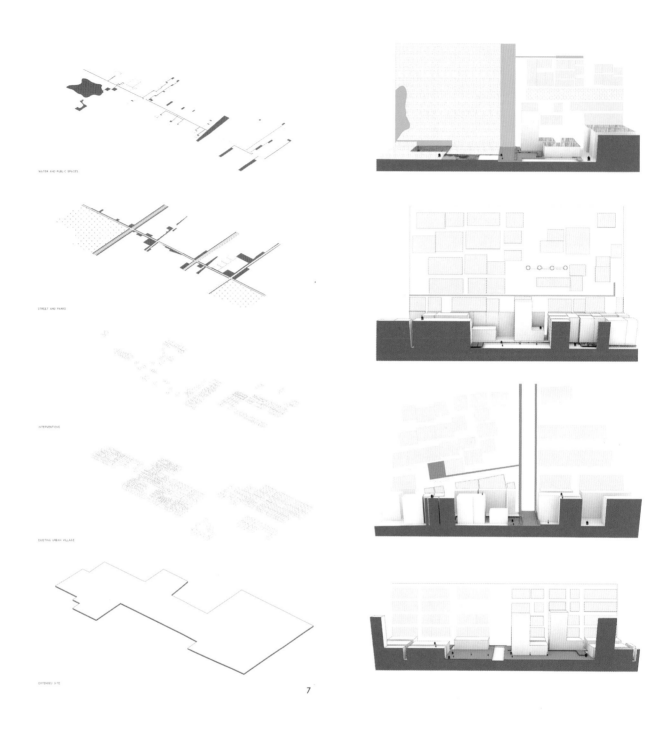

7

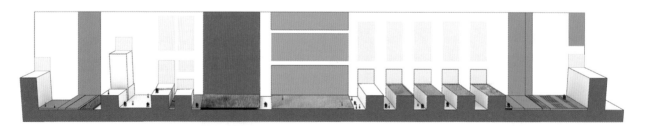

8

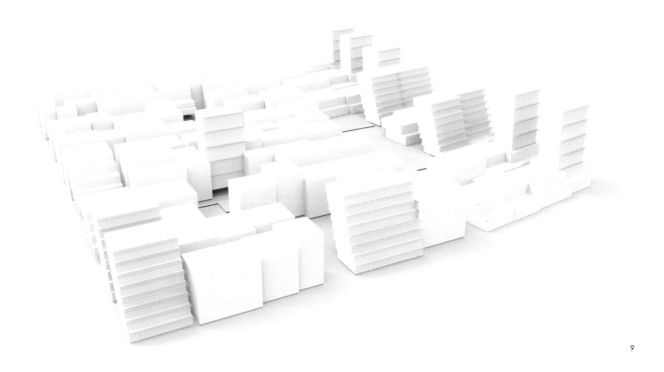

9

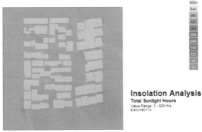
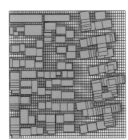
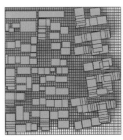

10

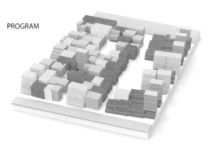

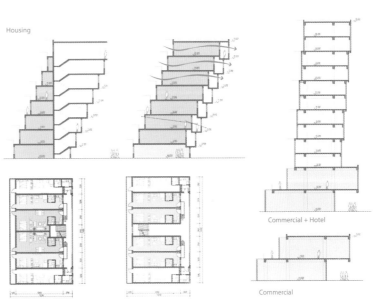

7 场地分层示意 Layers of Site
8 场地剖面示意 Sections
9 组团鸟瞰 Bird View of Cluster
10 组团风影响、日照影响 Wind and Solar Experiments
11 组团功能示意 Function
12 组团内户型 Housing Type

12

143

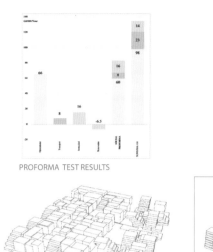

PROFORMA TEST RESULTS

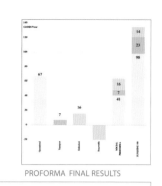

PROFORMA FINAL RESULTS

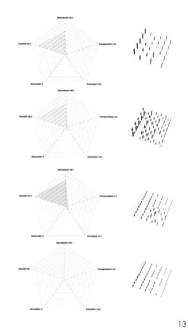

13

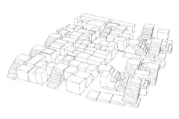

Summer Wind   Winter wind

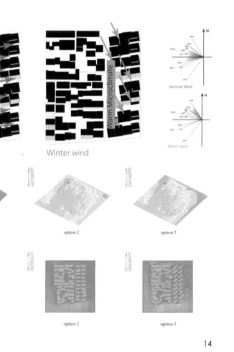

option 1   option 2   option 3
pressure simulation

option 1   option 2   option 3
velocity simulation

Phytoremediation

Algae Biofuel Shader

14   15

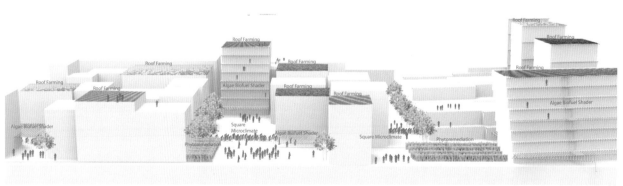

16

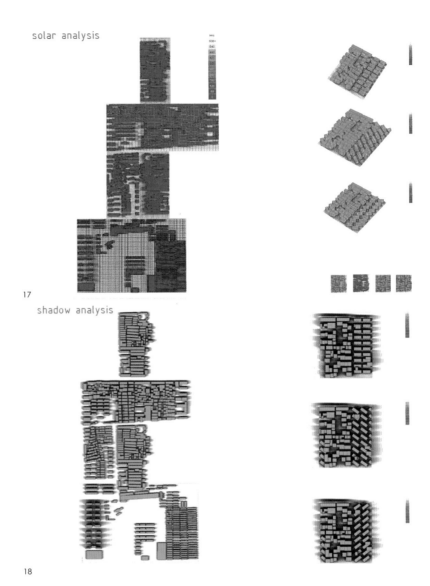

13 组团邻里模型能耗分析及筛选 Energy Analysis in the Neiborhood
14 组团内建筑朝向及风影响 Wind and Building Orientation
15 城市绿植 Green in City
16 城市绿植意象 Concept of Green in City
17 地段整体日照分析 Solar Analysis
18 地段整体阴影分析 Shades Analysis
19 场景意象 Perspective

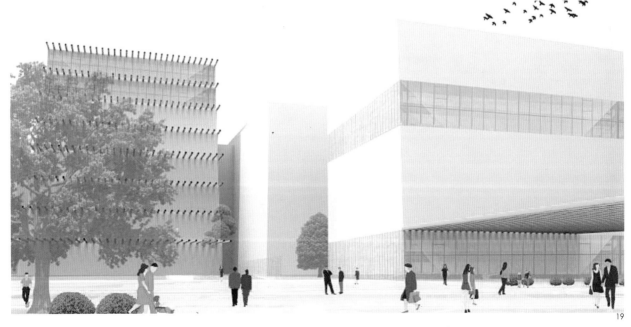

解
DISTRIBUTED MILLENNIUM

Chenxing Li, Meng Ren, Anson Steward, Linda Wu, Xu Zhang

restore water and green space systetms to create armatures of high quality shared space/ 恢复水道以及绿地系统以创造高品质的共享空间体系

integrate education and healthcare institutions into the neighborhood to promote lifelong learning/ 将教育与医疗设施整合入邻里住区以鼓励居民的全龄段学习

1 设计鸟瞰图 Bird View
2 设计任务 Design Task
3 设计目标 Design Goal

将现有邻里和教育等机构的转变为分散学习与健康的联结空间。
Transform existing neighborhood and institutions into networked spaces for distributed learning and health.

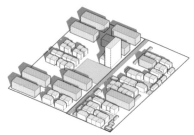

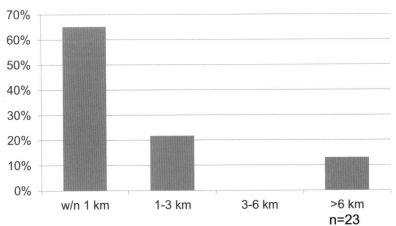

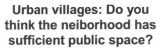

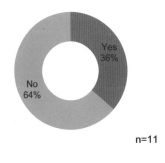

4 用地现状 Land Use
5 建筑密度 Density
6 交往距离统计 Communication Statistics
7 设计范围 Site
8 地铁系统 Metro

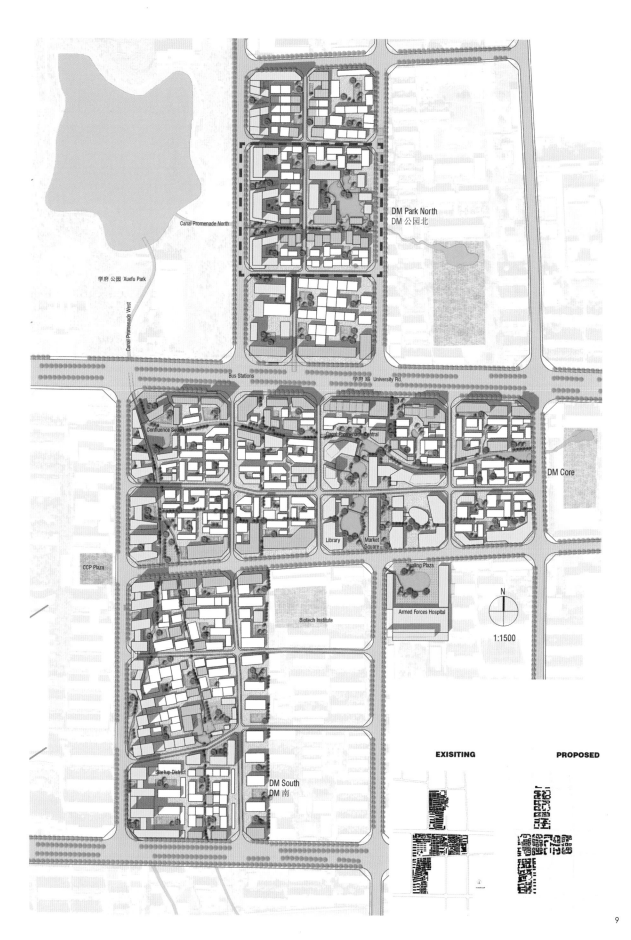

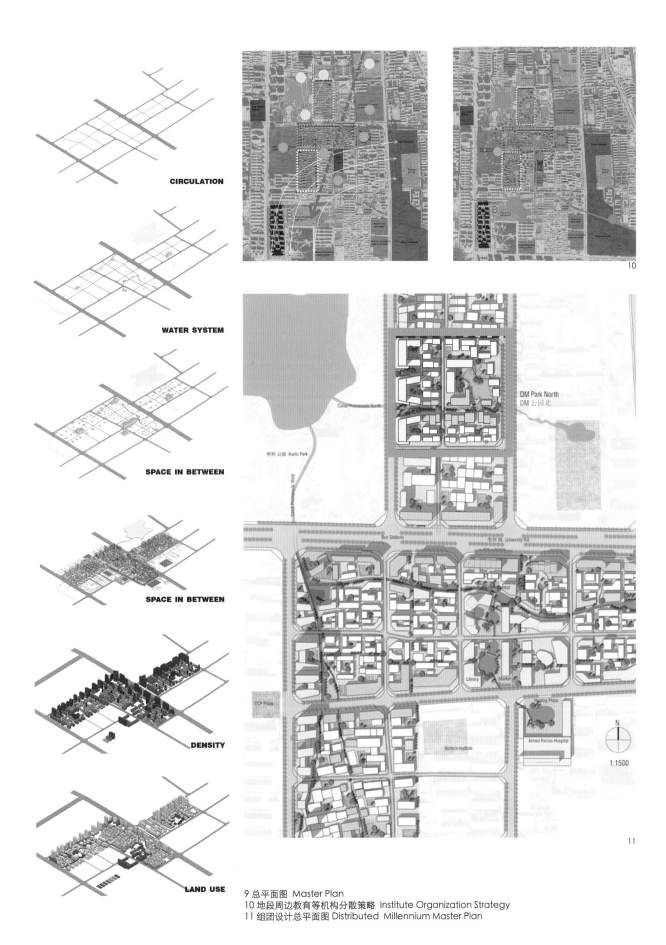

9 总平面图 Master Plan
10 地段周边教育等机构分散策略 Institute Organization Strategy
11 组团设计总平面图 Distributed Millennium Master Plan

PUBLIC ATMOSPHERE / 公共空间

View of the Sky Garden on Park Tower South. Urban agriculture and vertical gardens provide fresh food and opportunities for recreation.

View of Canal Promenade. Locals and visitors stroll this street and chat with each other.

View of the Elder Community, which has special amenities for retirees.

DESIGN EVOLUTION / 设计生成

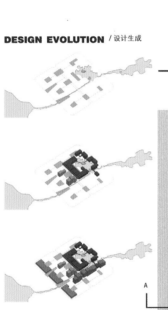

Section A-A 1:500

Cluster Plan 1:500

Total FAR / 总容积率:
3.6

Residential FAR / 住宅容积率:
2.8

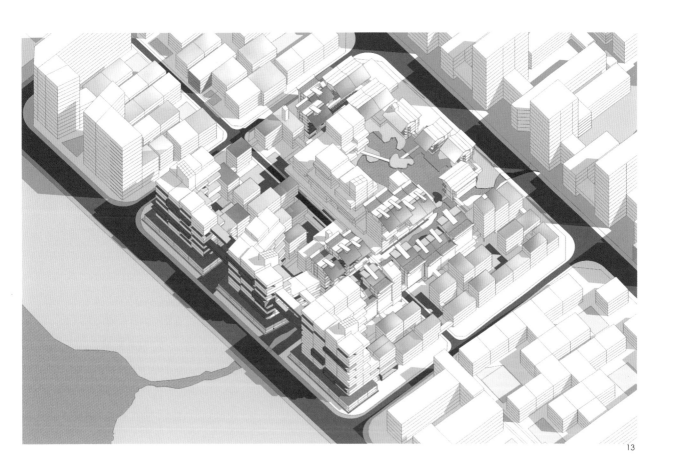

13

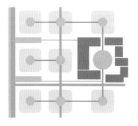

Walking Time to Transit / 步行换乘时间:
2-5 minutes

Parking Spaces / 停车:
625 underground 地下停车位
0.40 per household 每户

12 组团设计透视及意象 Cluster Design
13 建筑组团鸟瞰 Bird View
14 组团户型设计 Housing
15 组团内步行 / 停车流线 Circulation

14 15

151

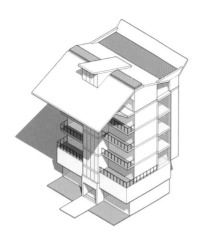
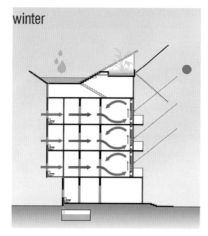
winter

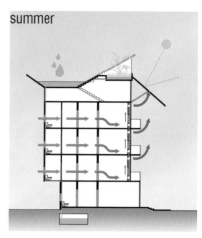
summer

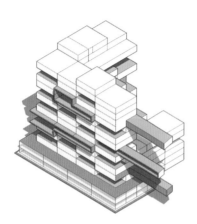
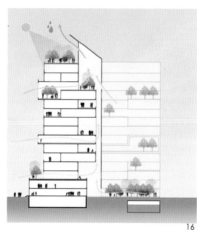

Neighborhood

1 Increase density/ 增加密度

Increase residents in this central, transit- and NMT- accessible neighborhood

2 Link to high quality transit/ 高可达性

Promote alternatives to private vehicle use

3 Promote environmentalism across activities/ 增强环境认同提高混合使用

Make environmental consciousness part of a desirable lifestyle by featuring it across spaces of living, work, and play

Block

4 Shape comfortable and active shared spaces/ 塑造舒适活力共享空间

Complement private micro-units with high-quality public spaces

5 Innovate linked energy systems/ 关联性能源体系

Share stack-effect chimneys and grey water systems between buildings

6 Re-localize agriculture/ 再现本地农业

Provide food, recreation and evapotranspirative cooling through gardens that evoke the areas agricultural heritage

Building

7 Promote micro-units/ 增加微住宅体

Reduce HVAC demands and match the economic and mobility needs of residents through smaller apartment areas (average 47 m²)

8 Design multi-functional building systems/ 多功能建筑体系

Photovoltaic panels can generate electricity and shade

9 Conserve water/ 涵养水源

Collect and store rainwater through rooftop cisterns

Average Building Height / 建筑均高
25 m

Average HH Annual Energy Use / 平均家庭能耗
70k MJ

Average Apt. Area / 平均住宅面积
47 m²

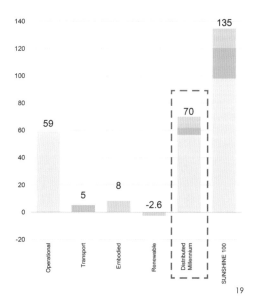

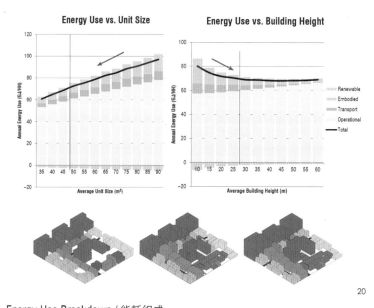

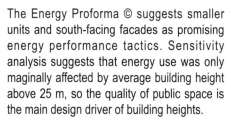

The Energy Proforma © suggests smaller units and south-facing facades as promising energy performance tactics. Sensitivity analysis suggests that energy use was only maginally affected by average building height above 25 m, so the quality of public space is the main design driver of building heights.

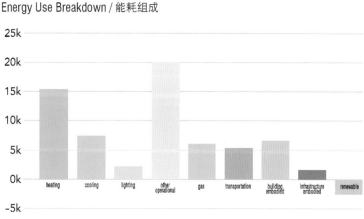

Energy Use Breakdown / 能耗组成

16 单体能源策略 Energy Features in Single Building
17 组团能源策略 Energy Features in Clusters
18 环境混合使用透视图 Perspective of Mixed-use
19 组团能耗敏感度与设计演进
Sensitive Analysis and Design Evolution
20 组团内能耗比较 Benchmark Comparison
21 组团内能耗组成 Energy Use Breakdown

原之线
URBAN TREADS
THREADSDISCOVERING THE JEWELS OF TAIYUAN

Allegra Fonda-Bondardi, David Zheyuan Xu, Kuan Butts, Rena Yang, Thi Tram

一个寓言：缀有珠宝的长袍
一个穷人在一个酒店帮助了一个富有的旅行者，为他的旅程提供了宝贵的帮助和指引。为了表示感谢，富有的旅行者秘密地将一个珍贵的宝石缝进了穷人长袍的衬里。这个穷人继续自己的旅程，许多年后，在他垂暮之时才发现了珠宝。

Parable of the Jewel-Lined Robe:
A poor man in an inn befriends a wealthy traveler, providing him invaluable help and directions on his journey. In thanks, the traveler secretly sews a valuable jewel into the lining of the poor man's robe. The poor man continues on his own journey, only to discover the jewels years later at the end of his life.

编制穿过庭院的线
我们在北张村邻里社区的调查试图通过对中国传统庭院形制的不同尺度的演绎将重要的公共设施——"珠宝"植入到现存的场地肌理中。这个充满活力和节能高效的城市肌理中的"线"包括人行小径、道路、生活用水系统以及实现废弃物再利用的商业区。庭院系统提供住房、高密度的商业空间，并与公园和绿色步道紧密相连。

Weaving threads through the courtyard
Our interventions in the Bei Zhang neighborhood of Taiyuan seek to weave the "jewels" of important public institutions into the existing fabric based on a multi-scalar interpretation of the traditional Chinese courtyard form. "Threads" of this vibrant and energy-efficient urban fabric include pedestrian paths, roadways, living-machine water systems, and waste-to-production commercial zones. The courtyard systems provide housing, high-density commercial space, and interconnectivity with parks and green pathways.

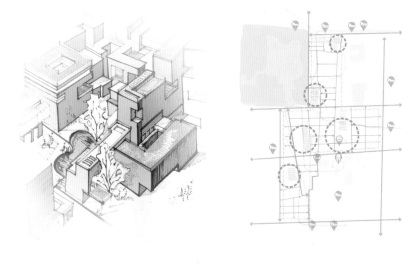

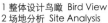

1 整体设计鸟瞰 Bird View
2 场地分析 Site Analysis
3 场地印象 Site View
4 总平面图 Master Plan

MASTER PLAN ELEMENTS

LAND USE PLAN

CIRCULATION PLAN

SPACE BETWEEN PLAN

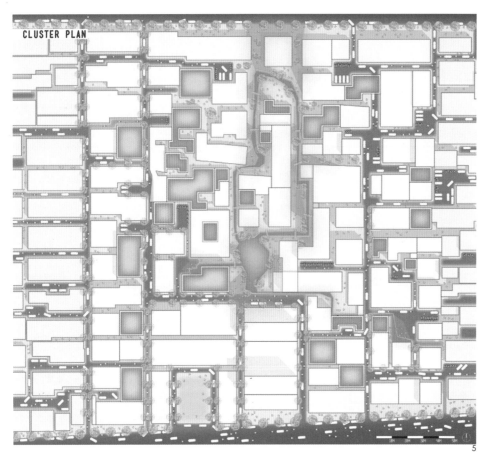

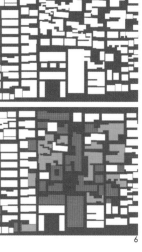
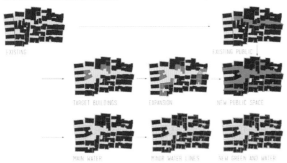
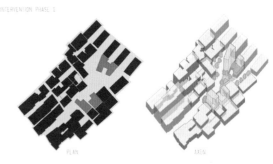

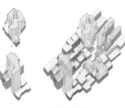

5 组团总平面图
Cluster Plan
6 组团位置选取及改动程度
Changes in Clusters
7 邻里介入策略
Neighborhood Intervention Strategy
8 建筑介入策略
Building Intervention Strategy
9 整体介入策略
Intervention Strategy
10 新邻里单元的建立
New Neighborhood Axon
11 剖面分析及剖面策略
Section and Strategy

NEW NEIGHBORHOOD AXON
1 TALL BRIDGE
2 ENCLOSED COURTYARD
3 TALL REACH OVER
4 STEPPED REACH OVER
5 STAND ALONE

NEW BUILDINGS WITH F.A.R

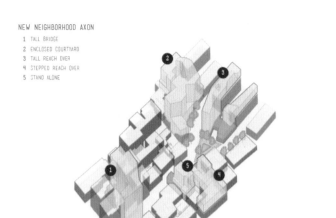
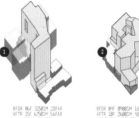

SECTIONS AND STRATEGIES

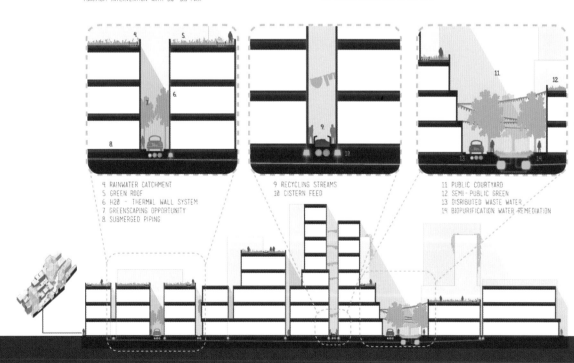

MAXIMUM INTERVENTION WITH 5.0 - 5.5 F.A.R.   NEW CLUSTER WITH GREEN ROOFS AND PARKS

1. TWO-LANE STREET
2. BOLLARDS
3. DISTRIBUTED PIPING

4 RAINWATER CATCHMENT
5 GREEN ROOF
6 H2O - THERMAL WALL SYSTEM
7 GREENSCAPING OPPORTUNITY
8 SUBMERGED PIPING

9 RECYCLING STREAMS
10 CISTERN FEED

11 PUBLIC COURTYARD
12 SEMI-PUBLIC GREEN
13 DISTRIBUTED WASTE WATER
14 BIOPURIFICATION WATER REMEDIATION

F.A.R. COMPARISON

PROPOSED

OPERATIONAL VS EMBODDIED ENERGY
(PROPOSED)

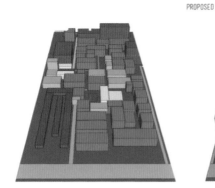
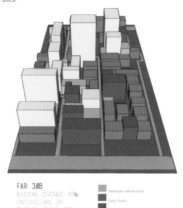
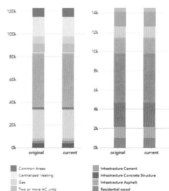

12

13

TOTAL ENERGY - USE COMPARISON

ENERGY - USE BY TYPE

EXISTING VS PROPOSED

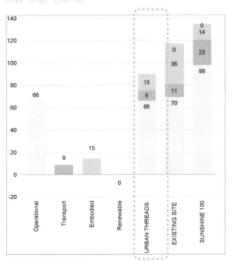
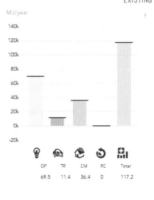
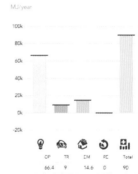
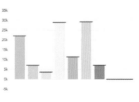
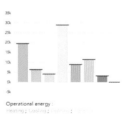

14

- PREDICTED 23.2% REDUCTION IN TOTAL ENERGY USAGE COMPARED TO EXISTING SITE CONDITIONS
- INCREASE IN FAR FROM 1.30 EXISTING TO 3.08 PROPOSED
- INCREASING DENSITY WHILE MAINTAINING TRADITIONAL COURTYARD URBAN FORM YIELDS ENERGY SAVINGS

12 能源评估 Energy Assesment
13 能源消耗预估与现状 Energy Consumption
14 地段总能耗现状 Total Energy
15 不同类型能耗现状与预期 Energy-Use by Type
16 建筑分析——置入庭院 Courtyard Schemes

15

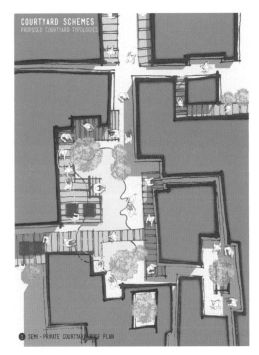

① SEMI-PRIVATE COURTYARD ROOF PLAN

COURTYARD SCHEMES
PROPOSED COURTYARD TYPOLOGIES

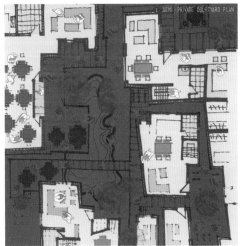

① SEMI-PRIVATE COURTYARD PLAN

① SEMI-PRIVATE COURTYARD

NARRATIVE STRATEGY FOR COURTYARD DESIGN

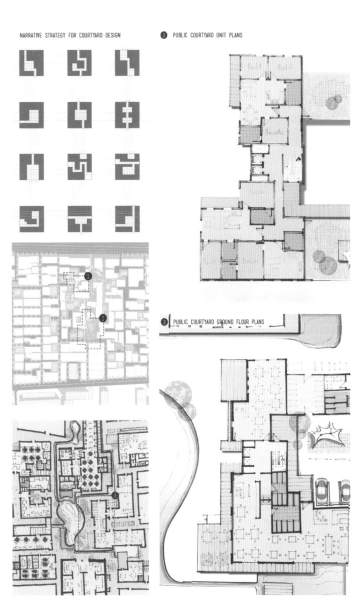

③ PUBLIC COURTYARD UNIT PLANS

③ PUBLIC COURTYARD GROUND FLOOR PLANS

② PUBLIC COURTYARD

17 组团内总体能源策略 Energy Strategy
18 组团内庭院空间意象 Perspective

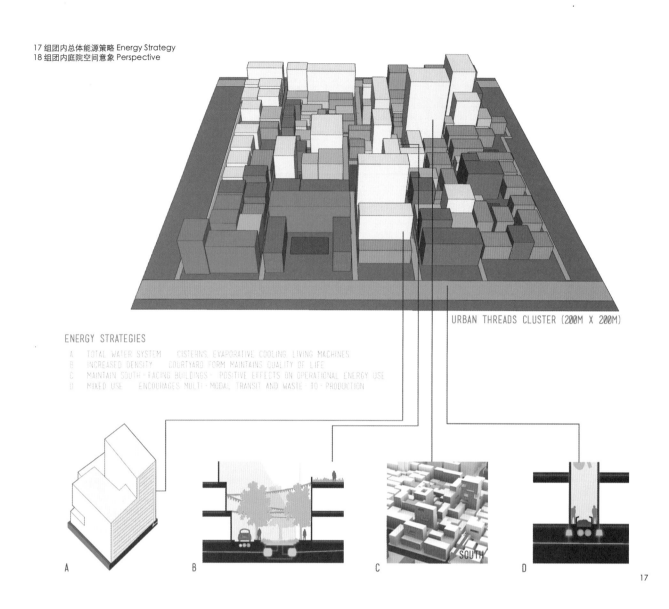

URBAN THREADS CLUSTER (200M X 200M)

ENERGY STRATEGIES
A TOTAL WATER SYSTEM   CISTERNS, EVAPORATIVE COOLING, LIVING MACHINES
B INCREASED DENSITY   COURTYARD FORM MAINTAINS QUALITY OF LIFE
C MAINTAIN SOUTH-FACING BUILDINGS - POSITIVE EFFECTS ON OPERATIONAL ENERGY USE
D MIXED USE   ENCOURAGES MULTI-MODAL TRANSIT AND WASTE-TO-PRODUCTION

17

PUBLIC COURTYARD + MARKET CONNECTION

18

韵转未来
FLEXIBLE FUTURES

Kara Elliott-Ortege, Ethan Lay-Sleeper, Boris Berndtson, Kun Qian, Jixiao Wang

1

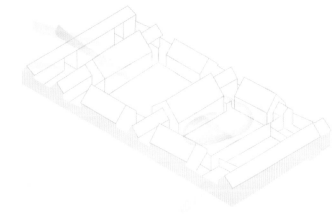

2

本方案围绕混合密度开发、多代社区、一生之宅、律动空间和穿梭流线展开设计。力求打造一个充满活力的、融合各个社会群体的可持续城市生活空间。

The project focusing on 5 perspectives, which are mixed-use density, multi-generational communities, aging in place, active urban open spaces and porosity. The oringal urban spatial design concept is derived from traditional courtyard housing, the project intend to create a three-dimensional courtyards which aims at maintain the local characters while satisfiy the living request of the residents.

3

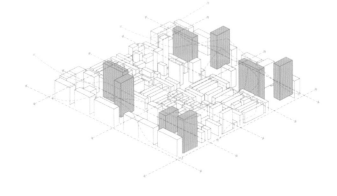

1 现场调研 Field Research
2 传统院落 Traditional Courtyard
3 三维院落 Three-dimensional Courtyard
4 密度 Density

4

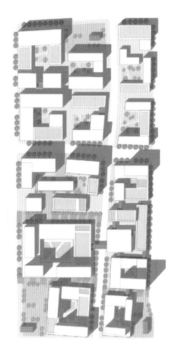

overall land use 土地利用规划

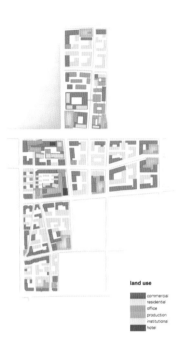

6

figure ground 图底关系

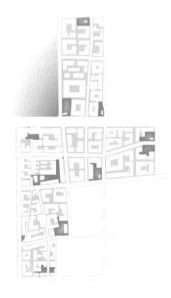

7

circulation + urban nodes 交通及区域节点

- - - vehicular roads 车行道
- - - restricted vehicular roads 区域车行道
——— pedestrian/bike only 自行车道 / 人行道

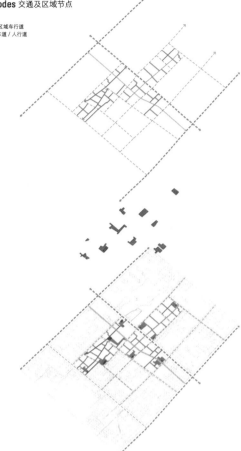

8

ground floor commercial use 底层商业

9

5 总图 Master Plan
6 土地利用规划 Land Use
7 图底关系 Figure Ground
8 交通及区域节点 Circulation + Urban Nodes
9 底层商业 Ground Floor Commercial Use

cluster 1: rail station + production
组团1：地铁及社区产业

on-site production and workshop spaces 提供场地内产业岗位
rail station plazas lined with commercial activity 地铁入口广场与商业并存
density towards the park and station 公园及地铁密度集中区

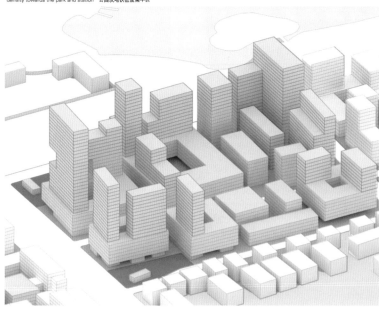

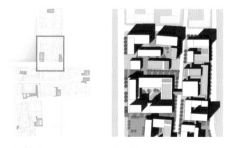

land use of cluster 土地使用分布

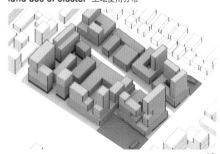

10

cluster 2: retail center
组团2：商业中心

urban courtyards lined with commercial activity 商业院落与商业活动并存
regional retail draw to the north, local retail activity within the plaza to the south 广场北部为区域性商业中心，南部为本地社区商业。
mixed use for office and residential 办公与居住之混合
density facing the park 密度向公园集中

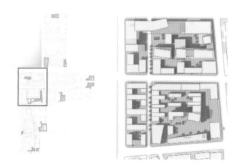

land use of retail cluster 商业组团土地使用分布

11

10 组团一：地铁及社区产业
Cluster 1: Rail Station + Production
11 组团二：商业中心
Cluster 2: Retail Center

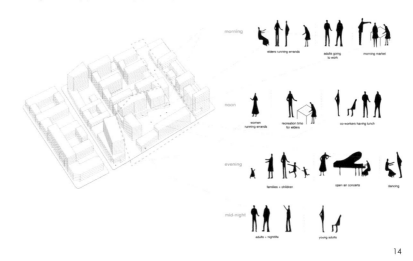

temporal use 短时使用

multi-generational population: demographic change within the open space throughout the day

12

14

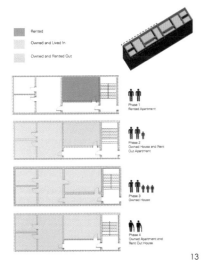

13

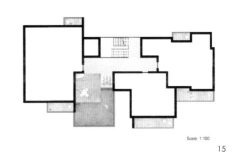

15

住宅规划策略

控制可支付性。调整住宅类型分布，促进场地内高收入人群、常驻户与短租户之间交流。

多代混合居住。提供多种户型灵活使用。促进多种住户入住：包括家庭、单身和老人。

灵活增收。灵活户型可根据不同时期家庭大小及资金需求改变使用模式。多变的居住方式使居民长期生活在社区成为可能。

垂直院落。垂直住宅中仍保有传统院落模式。密集住宅分布中增加透光及通风。

Housing Strategies

Maintain affordability. Mixed distribution of housing for both high and low income across the site catering for a variety of users: homeowners and renters.
Integrated generations. Providing housing to accommodate flexible living, encourage diverse occupants: families, singles, and elderly.
Flexibility to generate income. Units that can be adapted over time to respond to household size and income needs. Flexible living allows the resident the option to remain in the same community.
Vertical courtyards. Maintaining the idea of the traditional courtyard within vertical housing. Increases light and air ventilation within the dense urban form.

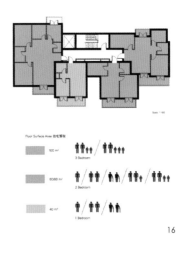

16

12 控制可支付性 Maintain Affordability
13 灵活增收 Flexibility to Generate Income
14 短时使用 Temporal Use
15 垂直院落 Vertical Courtyard
16 多代混合居住 Mixed in Generations

165

energy 能源策略及分析

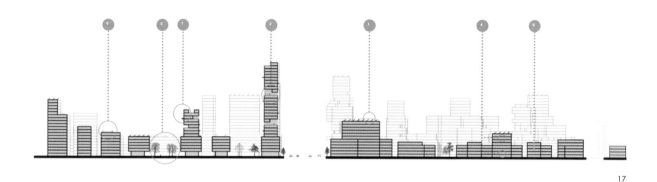

1. solar PV for hot water and energy generation   光伏板产能及热水
2. vertical ventilation for green space, light and air   垂直通风、透光、绿化空间
3. pedestrian streets for better circulation and air quality   人行道促进交通及通风
4. mixed-use communities to reduce travel & provide jobs   混合社区减少交通并增加就业
5. rain water collection   雨水收集
6. re-use of existing buildings for reduced embodied energy   原始建筑再利用降低建造能耗
7. south-facing facades for light exposure   增加南向立面透光性

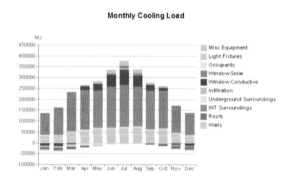

月制冷能耗

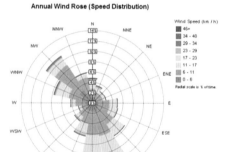

全年风速分布

月取暖能耗

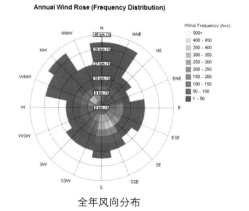

全年风向分布

17 能源策略 Energy Strategy
18 能源指标分析 Energy Criteria

166

低速风环境模拟

高速风环境模拟

19

total energy charts 能耗比较

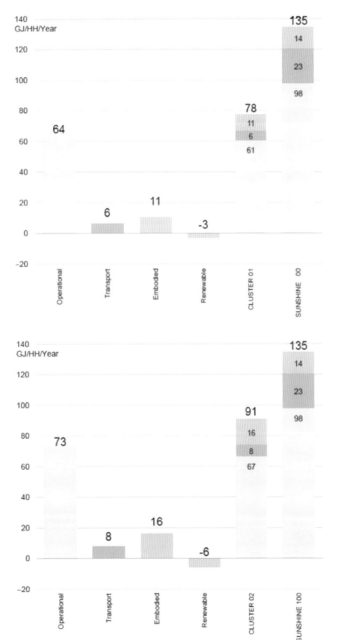

function: residence, production
主要功能：居住、社区产业

FAR 容积率：	5.21
avg building height 建筑均高：	34m
apt avg unit size 户均面积：	65m
green space ratio 绿地率：	7%
avg building footprint 均基底面积：	399m²
building function mix 功能混合率：	39%

function: residence, commercial
主要功能：居住、商业、办公、酒店、娱乐

FAR 容积率：	4.16
avg building height 建筑均高：	26m
apt avg unit size 户均面积：	75m
green space ratio 绿地率：	5%
building function mix 功能混合率：	70%

19 环境模拟 Environmental Simulation
20 能耗比较 Total Energy Charts

20

街道主义
STREETISM

Ma Yugang, Ge-Pei Meizi, Akhila Jambagi, David Stephan Jones, Rüdiger Schätzler

本方案旨在利用街道作为可持续能源系统的骨架,从地区活力、绿色节能和技术问题三方面进行详细设计。

The project concept is using the streets as skeleton for an energy efficient system. There are three elements closing interwoved with the streets: Vitality of the Village, Energy Issues and Practical Problems.

1 现场调研 Site Research
2 节点分析 Site Analysis

现有能源问题
基地现存室外庭院空间和屋顶空间缺乏利用、基础设施不安全、高能源消耗和街道立面隔热和玻璃状况差等问题。

能源原则
最大化能源再利用；增加绿化空间，改善城市小气候；保证居民对公共服务和基础设施的高度可达性；增加共享智能空间；提升公共空间质量，普及能源消耗知识；尽量采用翻新手段，避免推倒重建。

General Issues of Energy

Unutilized outdoor courtyard. Unsafe infrastructure. Unutilized rooftop area. Solar district heating system. High energy consumption. Facade in bad condition in term of insulation and glasing.

Energy Principle

Maximise renewable energy. Increase use of greenery to improve urban climate. Ensure local accessibility to a diverse range of goods and services. Optimise resources by implementing shared and smart systems. Promote use of public spaces and improve understanding of energy consumption. Saving embodied energy by preferring refurbishments to rebuilds wherever possible.

3

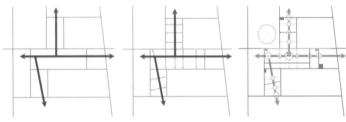

Step 1: Culture streets   Step 2: Branch streets   Step 4: Car ways

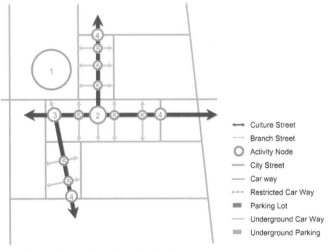

Step 3: A hierarchy of public activity nodes

3 概念生成 Concept
4 空间结构 Conceptual Structure

4

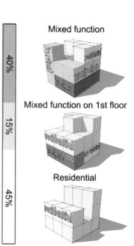

5 总平面图 Overall Plan
6 总体鸟瞰图 Overall Model
7 功能混合 Mixed use

高度
新建建筑多为5层以上,沿街和重要节点周边建筑较高,内部居住院落以3~5层为主。
流线
街道体系延伸至整个区域,步行为主,多元交通网络,采用道路下穿来联系割裂的场地。
街巷空间
街巷空间是公共最重要的外部公共空间,串连场地内外的重要公共空间和绿地空间。

Height
Buildings along culture streets and around activity nodes are higher. The internal courtyard is 3-5stories.
Circulation
Street system extends to the entire area. Streets are pedestrain-oriented and there is a diversified transportation network. Tunnels is used to connect the northern and southern part,
Overall Space Between
The street space is integrated within the open space system. Some streets are extended to connect a larger area. And the inside open space is connected to surrounding open space too.

Sky bridge
1-2 stories
3-5 stories
6-7 stories
8-10 stories
11-14 stories

8

Underground parking
Tunnel entrance
City main road
City secondary road
Inner car road
Culture street/
Restricted car road
Branch street
Main path
Tram
Tunnel
Main parking

9

Xue fu park
Main public space around
Space above tunnel
Street space
Green space
Extended street space

10

8 高度分析 Height Analysis
9 流线分析 Circulation Analysis
10 街巷空间分析 Overall Space Analysis

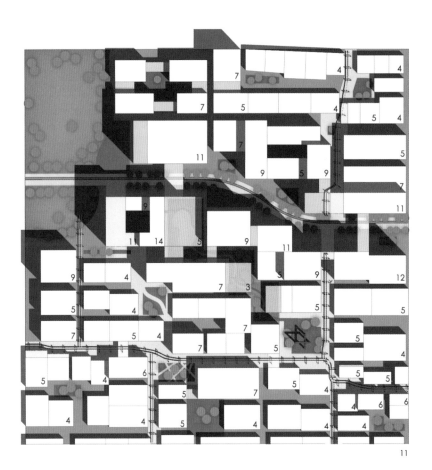

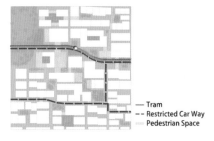

TO promote walking and public transportation
鼓励步行及公交出行

— Tram
-- Restricted Car Way
　 Pedestrian Space

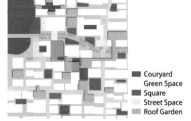

TO encourage outdoor public activity
倡导公共活动

■ Couryard
　 Green Space
■ Square
　 Street Space
　 Roof Garden

Culture Street: traditional village activity
文脉之街：传统乡村活动

City Street: new urban function
城市之街：新型城市功能

Sight Axis Street: node connection
视轴之街：联系各个节点

Street System: a comprehensive system
街巷体系

12

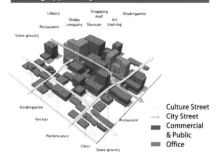

TO provide local accessibility to goods and working opportunity 功能混合

— Culture Street
→ City Street
■ Commercial & Public
■ Office

TO enhance natural ventilation and lighting
加强自然通风与采光

Buildings along the city street are higher to create micro-climate so as to enhance natural ventilation. 创造微气候，促进自然通风。

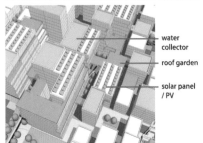

TO make use of renewable energy
利用可再生能源

water collector
roof garden
solar panel / PV

Different spatial experience, various activities
多样的空间体验，丰富的公共活动

11 组团平面图 Cluster Plan
12 组团分析图 Cluster Analysis

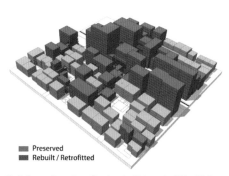

- Preserved
- Rebuilt / Retrofitted

Buildings along the city street will be rebuilt for higher FAR. Some buildings along culture streets will be rebuilt for special functions.
沿城市之街的建筑和部分沿文脉之街的建筑将会重建。

- Residence
- Office Working
- Commercial / Public Facility

There will be a mixture of residence, offices and commercial along the city street. Commercial and public facilities will also be on the 1st and 2nd floors of buildings along culture streets.
居住、工作场所、商业和公共服务设施沿街巷高度混合。

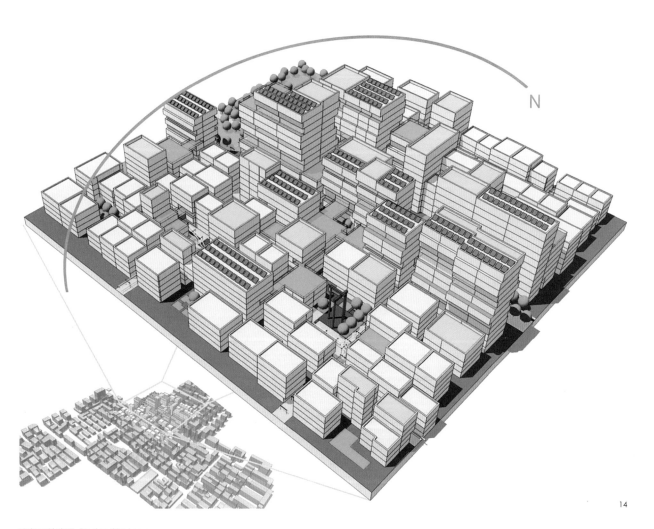

13 组团策略图 Design Strategy
14 组团模型 Cluster Model

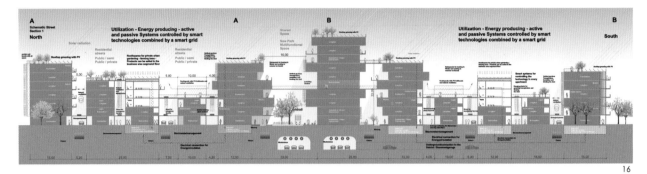

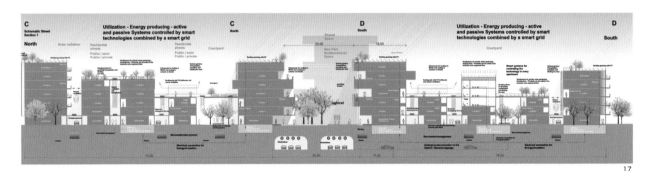

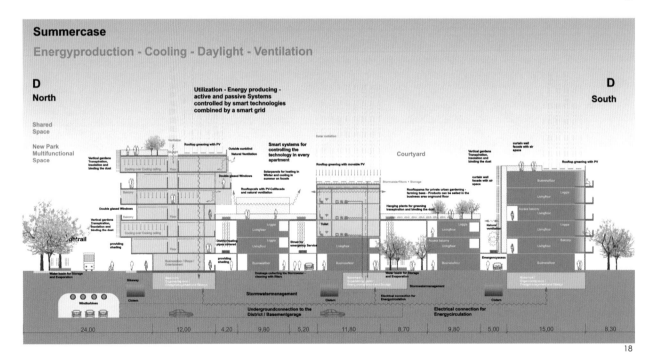

15 剖面位置 Section Position
16 剖面 A-A B-B Section A-A B-B
17 剖面 C-C D-D Section C-C D-D
18 剖面 D-D Section D-D

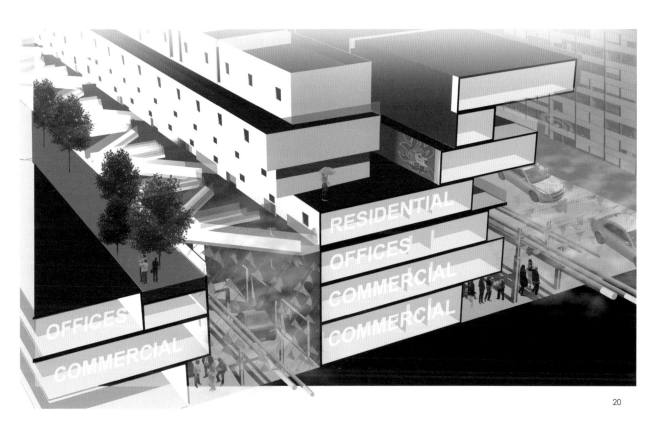

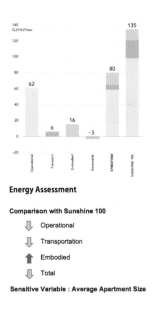

Energy Assessment

Comparison with Sunshine 100

⬇ Operational
⬇ Transportation
⬆ Embodied
⬇ Total

Sensitive Variable : Average Apartment Size

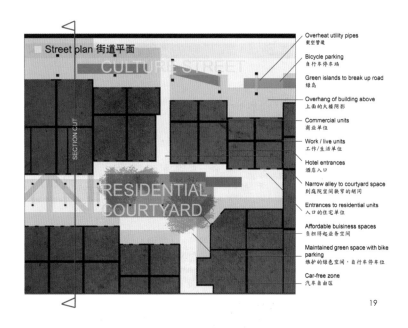

19 街道平面 Streets Plan
20 节点可视化 Building Visual

路径
LIVE/ WORK/ PLAY PATHS
F. Jinglin, M.A. Kramer, P.E. Little, A.V. Marchetti

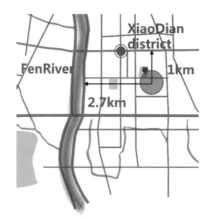

Our site located in the Xiaodian district, which is one of the four centers of Taiyuan. The site is 2.7km from Fen River and 1km from ChangFeng Steert which is the one of axis of the city.

本方案旨在营造社会交往的空间，通过功能混合融合不同社会群体，并打造新的休闲项目。通过建立绿道系统衔接学校和广场空间吸引人群进驻。

The project aiming at provide opportunities for socia interaction by applying mixed-use strategies while create new leisure activities. The major spacial intervention focusing on establishing a green path system through connecting the schools and plazas. The site will obtain a safer and greener enviroment through a gradual process.

解决策略

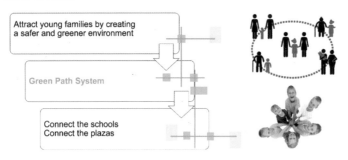

1

1 基地区位 Location
2 现状问题 Problems
3 目标提出 Objective

2

Concept 思考

Objective 目标

3

Buildings

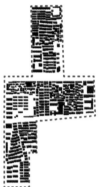

Streets

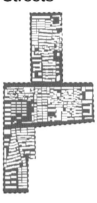

Open spaces

Pedestrain paths

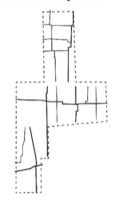

Commercial

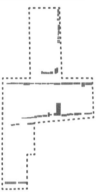

4

4 现状分析 Exsiting Site
5 方案生成过程 Progress

Progress
方案生成过程

Step 1:
Locate the surrounding schools

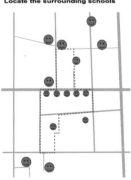

Step 2:
To seek the direct time-saving way to combine all the direction of flow

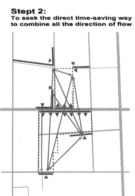

Stept 3:
Combine the exsiting texture of street and rule of road to find the most appropriate path

Stept 4:
Add plazas in the intersection of thepath

5

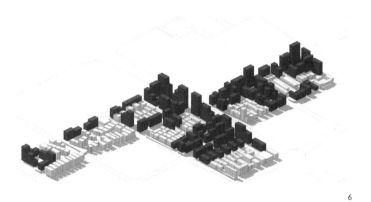

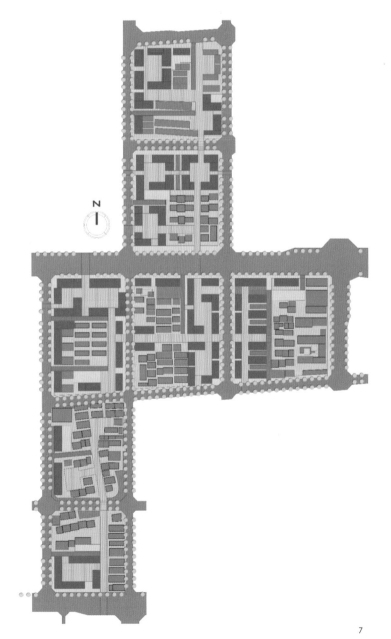

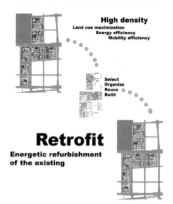
New construction
High density
Land use maximization
Energy efficiency
Mobility efficiency

Select
Organize
Reuse
Build

Retrofit
Energetic refurbishment of the existing

Policentrism
The plaza system
Socialization
Commerce
Access to the sun

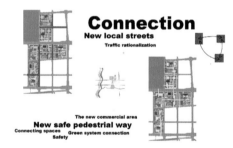
Connection
New local streets
Traffic rationalization

The new commercial area
New safe pedestrian way
Connecting spaces   Green system connection
Safety

Exclusion
Time scheduled
Area with traffic No parking
Only residents and deliveries
restriction Traffic reduction

Dead end street
The acces to the core of the site
Parking-Emergencies

Mixed use
Energy waste reduction

Buildings use
Mixed use prevalence residential
Mixed use prevalence residential
Residential destination

6 总体鸟瞰图 Bird View
7 总平面图 Overall Plan
8 策略分析 Strategy Analysis

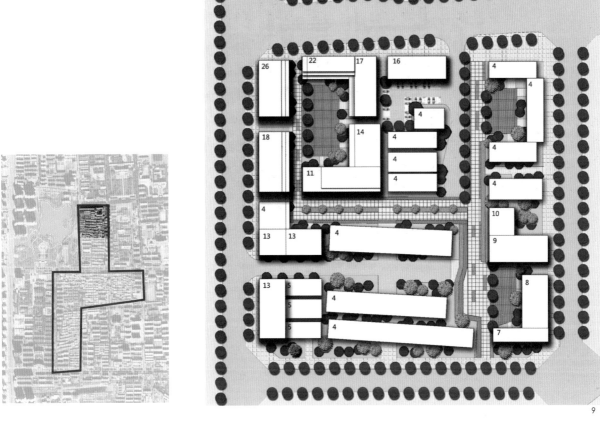

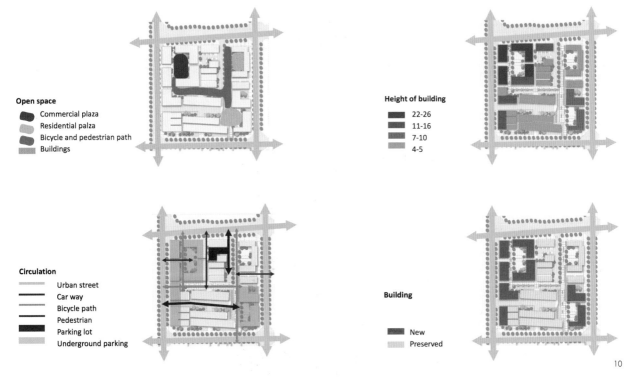

9 节点平面放大 Cluster Plan
10 节点分析 Cluster Analysis

179

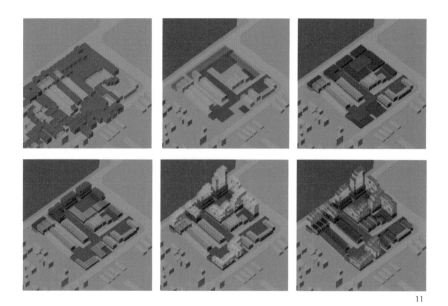

管理能源消耗——参数与影响分析

南向建筑立面比率增加2%，终期建筑运行能耗将会降低3.5%。建筑功能混合度23%，更高的建筑功能的组合导致更高的人均运行能耗。

Operational Energy Analysis

Southern-facing wall ratio: 32%, increasing the ratio by 2% reduced the operational energy by 3.5% in the final design step
Building function mix: 23%, the higher the building function mix, the higher the per HH operational energy since families tend to spend more time in their homes.

11 节点空间构成 Cluster Design Process
12 效果图1 Perspective 1
13 效果图2 Perspective 2
14 节能设施 Energy Strategy
15 指标分析 Criteria Analysis

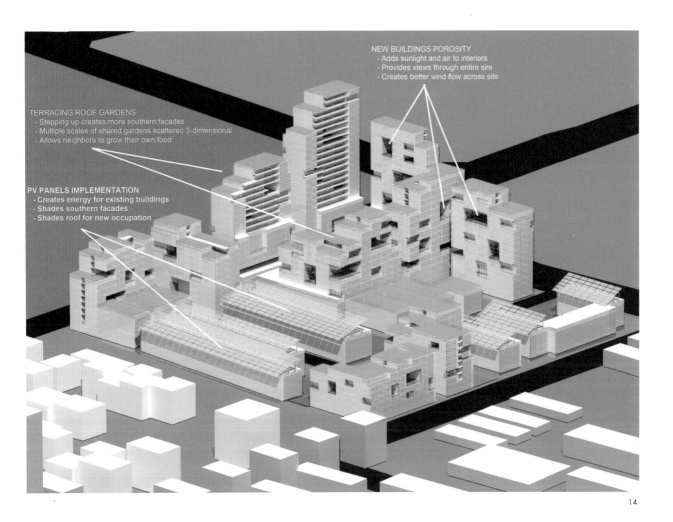

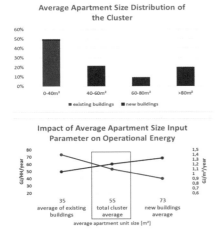

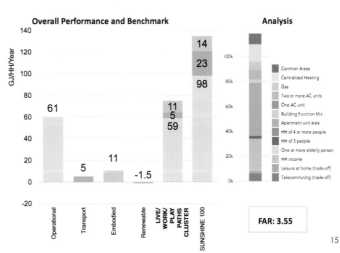

致谢
ACKNOWLEDGEMENTS

感谢清华大学建筑学院为本书出版提供的大力支持，感谢所有在近 10 年为清华 -MIT 城市设计联合课程中付出辛勤劳动的学者和同仁。同时还要感谢参加联合课程的各位同学，他们为本书提供了丰富的材料和精彩的作品。感谢清华大学建筑学院的牛泽文、袁周、谢湘雅、李旻华、陈恺、闫博、王祎、赵慧娟、尹子潇、张乃冰等同学为本书整理与排版做出的大量艰苦的工作。最后，我们还要感谢中国建筑工业出版社为本书的出版所做的一切。

We gratefully acknowledge the strong support from School of Architecture of Tsinghua University for this publication. And we would like to express our gratitude to the scholars and colleagues for their hard work in this THU-MIT joint studio of urban design in the past decade. Meanwhile, we deeply appreciate all the students working in this joint studio, who have provided rich materials and excellent works for this book, and also Niu Zewen, Yuan Zhou, Xie Xiangya, Li Minhua, Chen Kai, Yan Bo, Wang Yi, Zhao Huijuan, Yin Zixiao, and Zhang Naibing from School of Architecture of Tsinghua University, who have done a lot of hard work in editing and layout. Finally, we want to thank China Architure & Building Press for everything they have done in this publication.